Mathopedia
The Omni Guide to Math for Everyone

Edmund Ferrell

Omni Books
Greensboro, North Carolina

To my parents, Henry and Libby Ferrell

Editor: Eddie Huffman
Acquisitions editor: Stephen Levy
Editorial consultant: Janice Richardson
Cover design: Dwayne Flinchum
Interior design: Janice Fary
Typesetting: Terry Cash

Printed in the United States of America.

10 9 8 7 6 5 4 3 2 1

Library of Congress Cataloging-in-Publication data
Ferrell, Edmund.
    Mathopedia : the Omni guide to math for everyone / Edmund Ferrell.
        p. cm.
    Includes index.
    ISBN 0-87455-297-4
    1. Mathematics—Popular works.   I. Title.
QA93.F47   1993
510—dc20
                                  93-39739
                                      CIP

Omni is a registered trademark of Omni Publications International Ltd. Omni Books, 324 West Wendover Avenue, Suite 200, Greensboro, North Carolina 27408, is a General Media International Company.

Contents

Acknowledgments

Many people helped make *Mathopedia,* but nobody helped more than publisher Stephen Levy, and I thank him for that. A special thanks goes to Eddie Huffman and Terry Cash for literally putting the book together and making it better. Thanks also to Janice Richardson for her considered opinion, advice, and corrections. Final thanks to Keith Ferrell and the whole crew of the *Lucky Jack.*

Instructions for the Math Wary

Math is a science, a tool, an art, a game. Yet very often just the mention of the subject rings our Pavlovian trauma bell. Uncomfortable memories from math classes still haunt many of us today, and when we come face-to-face with a mathematical idea in the real world we freeze. We say we can't "do" math, and we can't. We never liked math, so we don't.

Mathematics is a lot more than just calculating, but unfortunately most of our experience in school emphasized only rote problem-solving and testing, testing, testing. Scant attention was paid to the lives of the people that "do" math, and even less attention was paid to the exciting life of math itself. So all we remember now are word problems and algebra, and we can't shake our fear of being called to the blackboard.

Our emotional defenses may go code red when we're forced to calculate, but we happily, if unconsciously, put math to work in our art, games, and checkbooks. There's a fascination and appreciation for this subject that is, in many ways, built into us. It tugs at us in our science fiction, confronts us on April 15th, and lies mysteriously beneath our fascination with lasers, warp speed, and worm holes. So what happened?

Early thinkers and doers didn't create mathematics so teachers could get chalky fingers. Early math did things. It predicted the eclipses and the seasons. A certain knowledge of math was a crucial navigational skill. Math built strong walls to protect young civilizations and helped an agricultural people increase their harvests to the point that there was some left over for the tax man. There was math to be found in painting and sculpture, music and games.

There still is. Whether you're good at math or not so good, whether you think you'd prefer a stick in the eye to math, math is everywhere. Computer graphics enhance our movies, digital CDs replace records, cellular telephones put our conversations on the road, and space flight takes our fellow humans— and our imaginations—skyward. All those things are steeped in mathematics. What are your chances at winning the lottery, or beating the house at the craps table? Speaking of craps tables, once a year your income taxes plunge you into the world of mathematics, if only basic calculations (hopefully not involving negative numbers).

The story of mathematics includes lots of interesting characters, practical problems, and serious arguments. The story takes place everywhere, including the inside of a lecture hall and in the letters mailed between gamblers. Maybe by looking at some of the things that mathematics did and the people who did them we can gain a better appreciation for the mathematics that is in and around us in everything we do.

The Beginning of a Long Story

T he first things that mathematics did for people—the very first things it did—it is still doing. Today we count and measure things easily, and these are among the first skills we teach our children. It's impossible to think of our world without the ancient arts of counting and measuring—and they are ancient indeed. There is evidence that people counted as long as 50,000 years ago.

There is little hope of ever learning anything about the specific individuals involved in the earliest development of counting and measuring. We have few reliable facts about even the most well-known mathematicians like Pythagoras, and he lived less than 3,000 years ago. For all we know, the very beginning came from the telling of a fish story.

Crouching by the flames of an ancient fire pit, an emerging human, probably hungry, remembers a meal caught on a bone hook. Grunting and boasting to his neighbors, this ancient angler stretches out both hands to show "how big." The fire now warms the hands that built it, and those same hands tell the tale that started math. Maybe.

Lots of observations of size and quantity were made which, over time, gave rise to mathematics. Math met people's growing need for a grammar to accompany the observations they were making—observations that could be painfully easy to make. An empty belly teaches better than a blackboard, and it doesn't take a Neolithic Einstein to see others in the tribe grabbing more than their share of roasted deer loin. But how much more?

The early story of mathematics is the story of human activity. Never mind how big your fish was; once people started counting and measuring, they would soon use these skills to build temples, survey land, mine metals. Far beyond the scope of any written records, early activities demanded that people answer the first mathematical questions: how much and how many?

Count On Me

Like the earliest counters, children of today still answer the question "how many?" with their fingers. Fingers are perfect to count with unless you have more than 10 items. Of course, you can easily count to 20 if you use both fingers and toes. In many cultures toe-counting is considered impolite, but in any culture you'll find some form of finger-counting.

Some people learn to count by first making a fist and then uncurling each finger as they count one, two, three. Another style starts with all fingers spread, counting each one as it is folded towards their palm or thumb. There are lots of different ways to use the fingers to represent numbers.

Many cultures, especially in the East, have taken finger-counting to a very high degree. Japanese and Korean finger-counters can perform difficult calculations with their systems. It's no longer a child's game; experienced finger-counters are employing a powerful and accurate method of computation.

Accurate or not, in the world of pocket calculators finger-counting seems to be a mere curiosity, but maybe there's more to it. Some studies indicate that exposure to serious finger-counting can increase a student's interest in math—even raise test scores. Not to mention how helpful it can be when your calculator batteries die.

TRY'EM—The ancient art of finger-counting

Try your hands at the ancient art of finger math. Think of your fingers as the numbered fingers in the diagram.

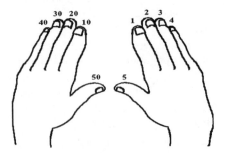

Place your hands on a desktop or on your thighs. Keep all fingers off the surface, and count the number represented by the corresponding finger or combination of fingers that is "pressed." Since all fingers are lifted, you start at zero. Counting one through ten would be as follows: press finger 1, say "one," then lift. Press finger 2, say "two," then lift. Press finger 3, and so on. Once the thumb is pressed for "five," keep it pressed while repeating the pressing of fingers 1 through 4, but counting "six, seven, eight." When the pinkie and thumb are pressed, say "nine," lift all your right-hand fingers, press your first finger on your left hand, and say "ten." Continue pressing the 10-finger while counting one through nine again with your right hand, counting 11, 12, until the first left-hand finger and the thumb and pinkie of the right hand are pressed for 19. When your thumbs and pinkies on both hands are pressed, you'll be counting 99.

To add numbers, just register the first number by pressing the appropriate fingers; for 47 plus 29, start by pressing 47. You will have your left pinkie and right thumb and right middle finger pressed. Then, without lifting your fingers, count 29. Read the final fingering: left thumb and middle finger for 70, right thumb and first finger for 6. The answer is 76.

To subtract, register the larger number on your hands, then count down the smaller number and read the answer off your fingers.

Multiplication is just adding over and over again. Using more sophisticated methods, finger math experts can reliably handle numbers into the thousands. The Venerable Bede introduced a method of finger-counting that was in use in Europe until recently. Now go practice by reconciling your checkbook using finger math.

Some people count on more than just fingers and toes. Near the coast of Australia, the Torres Strait Islanders counted on most of their body parts even as late as 100 years ago. Island clothing exposes more body parts to count, and by using wrists, elbows, chest, and ankles (among other parts), these Pacific mathematicians count by pointing at each body part in turn. The islanders end with the right little toe, number 33. A close-knit nudist colony might use this method to count even higher.

So finger-counting is in, even if you're just signaling "two outs left" to the outfielders, or indicating how many tokens you want in a noisy subway station. If you really want to see some high-level finger math, watch the traders on the stock market floor.

Measure For Measure

Lacking rulers and yardsticks, early people measured things with what they had on hand. Literally. Fingers were for counting, and hands were for measuring. Probably the oldest unit of measure is the cubit, which was used in early Egypt and Babylon. Cubits are created by human hands, but without the accordion-playing margin of error (or deception) that plagues the fisherman's method. It is simply the distance from the tip of the outstretched middle finger to the elbow. *Cubit* is Latin for elbow.

Of course, not everyone's cubit is the same length, but my cubit will remain essentially the same every time it's measured, at least after I've reached official size and weight. Someone's cubit had to be the standard, and, since early civilizations clumped around kings and pharaohs, debate was shortened over just whose cubit to use. Imagine the size of the pyramids if ancient engineers had taken their cubit measure off an emperor as big as Patrick Ewing!

Measurements were necessary in all sorts of human activity, like farming, mining, and sailing. Even today when people engage in these activities they use words that reflect the early connection between hands and measurement. Horses are still said to stand so many hands high.

To find the depth of the water, a sailor might drop a weighted line into the water. When it hit bottom, the tar would hold the line tight in one hand and loose in the other. By straightening out both arms and then bringing them together, the line would be measured in terms of arm-lengths. The sailor would know how deep the water was, and the line would be in shipshape coils. Would you be surprised to learn that *fathom* means outstretched arms?

Counting and measuring were going along just fine using only hands and fingers, but there were problems. There are more limitations to body counting and measuring than just social etiquette. While adequate for indicating amounts, even the sophisticated methods of finger-counting do not make for lasting records.

Sticks and Stones

More permanent counting records were needed, and again people chose what they found at hand. Sticks were especially cheap and available. A shepherd might herd two dozen sheep using a staff cut with 24 notches. If the wolves go hungry and no sheep wander off unnoticed, there will be one sheep for every notch on the staff come nightfall. A really long staff might indicate a very prosperous shepherd.

Special, smaller sticks were sometimes used to record debts. The notches were cut so that when the stick was split vertically, two "copies" were created, one for the person who loaned the money and one for the debtor. When the time to settle the debt came, the two halves of the stick were compared and a square deal was assured.

This simple system remained popular for a very long time. From the Latin word for stick they were called tally sticks, and were even used by the British Treasury Department. In fact, the British used a version of stick bookkeeping even into the early 1800s. Once they finally became obsolete for bookkeeping, the old tally sticks were used to heat the Houses of Parliament. This use of tally sticks proved to be dangerous, and was probably cause of a fire that leveled the Parliament buildings.

The Chinese also used sticks, but they used them in tandem with a counting board. The board was marked off into squares onto which four-inch sticks were placed to represent values. The counting board wasn't used to calculate, but for the temporary storage of numerical information. Empty squares with no match sticks indicated zero for that square's value. In this way, the Chinese system foreshadowed the introduction of zero and the place-value system.

Pebbles were also popular counting units, so popular that we derive many mathematical words from *calculus,* the Latin word for pebble. Even people who couldn't strictly count could match up one pebble for each goat, one pebble for each ram. A clay pot holding one pebble for each animal could be given to a trustworthy but non-counting servant. After a day of grazing the flock, the servant would use the pot to count the animals. While penning the animals at the end of the day, the servant would drop one pebble into the pot as each animal entered the pen. If there were pebbles left, an animal was missing.

Great Tool—The Abacus

Pebbles in a pouch were good, but pebbles on a rod were even better. For one thing, they wouldn't get lost. The modern form of the abacus was introduced in the 16th century, but some form was used at least by 500 BC. Even today the abacus is used in Eastern Europe and Asia.

The exact route of the abacus through history is hard to reconstruct, but it probably came out of Central Asia, moved west to Europe, and then east to China. Eventually the Europeans would come to prefer writing out their calculations, but many other countries, China chief among them, continued to work with the abacus. It was the cornerstone of ancient computation, and for thousands of years no merchant was without one.

In 1593, the modern abacus gained even more popularity in China when Ch'êng Taí-wei published *Suan-pan Computation. Suan-pan* means counting tray, and a modern Chinese counting tray is shown here.

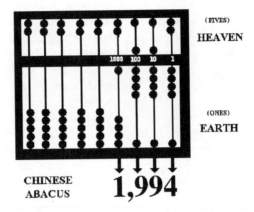

The top of the Chinese abacus is referred to as Heaven and the bottom as Earth. The top portion of the rods hold two beads which are used to indicate units of five, but they can also be used to perform base 2 arithmetic. The rods increase in value from right to left—one's place, ten's place—and so forth, just as our number system does. Sometimes the farthest two rods are used to count fractions.

Through much of the history of calculating people used an abacus. The abacus still enjoys popularity today, and they are still cheap, reliable, and easy to use. Even today an abacus expert can sometimes rival the speed of calculations by a hand-held calculator. Experienced fingers can make the beads on the abacus fly into position must faster than fat fingers can punch tiny buttons.

Now pebbles were really working out. Merchants, engineers, and government officials all had a tool they could count on thanks to the small pebble. Even before the abacus, people used stones to count things besides loaves of bread or goats in the field. Stones were useful for counting really big things, like the stars and the planets. It took some pretty big stones, but long ago people used them to measure time.

'Tis the Season

I n a flashback scene in Woody Allen's film "Annie Hall," a 10-year old Woody becomes so distraught that his mother takes him to the psychiatrist. The young boy is depressed, he complains to the doctor, because he read that the universe is expanding continuously, and at an incredible rate. At this point his mother shakes him by the shoulders and demands, "What is that your business?" Well, what is that your business?

At one time it was everybody's business. More than the depth of the water and more than the number of animals in the flock needed measuring in antiquity. Nomadic hunter-gatherers noticed the stars rising later each night of the year. Certain constellations appeared at the start of the different seasons, and so people began to *measure* the seasons.

By recording strokes on a clay tablet, one could recall how many days passed since the last flood. The number of days between moon rises could be grouped together, just as the number of moon rises from planting season to planting season could be grouped together. The seasons, like the lambs in the field, could be measured, and a permanent record could be produced.

At first it was a necessity. The better they became at mathematical stargazing, the surer they were to get to the fields when the bushes would be hanging with ripe berries; the more certain they would be of returning to the brackish mud flats when the shellfish were most abundant; the more precisely they would reckon the time the storms came down. As the nomads began to settle the fertile cradles of the world, math helped build straight walls to protect these young civilizations.

More exact mathematical methods were required for the more complicated jobs required of an advancing civilization. Fertile rivers flood their banks and enrich the land—or wipe out a settlement—depending in part on how much math is known in the settlement. Berry-picking season is easy to predict compared to producing an accurate timetable to use for lambing, sowing, and reaping.

An accurate measure of time required more memory than many could bring to the problem. The movements of the moon and sun could be charted by arranging stones in lines that pointed to the place certain constellations would rise from. Stone circles were arranged to mark the important movements of the moon and sun. The sun would reach a highest and a lowest point in the sky over the period of a year, and so giant stones were set up to sight the celestial events.

Stone Calendars, Candles, and Water Clocks

People began constructing Stonehenge around 1900 B.C., but other, simpler stone observatories had already been created. Most of these ancient arrangements provide for predicting the solstices and equinoxes, the longest and shortest days of the year. In this way, rock circles became the first observatories and the first calendars.

People needed to measure shorter lengths of time than the seasons of the year. More than 4,000 years ago, probably in sunny Egypt, people started measuring the length of the day with sticks. A straight stick thrust into the ground created a shadow that moved with the sun. Giant sundials were popular thousands of years ago. One sundial in Jaipur, India, had a gnomon that was more than 100 feet high. The shadow cast by this sun clock traveled over an acre throughout the day, and moved about a foot every five minutes.

The stick, called the *gnomon*, would be put inside a bowl around 700 B.C., and markings along the rim of the bowl were positioned to mark segments of the day. When the gnomon was tilted and pointed due north, the shadow of the bowl sundial would remain the same length as it moved around the rim of the bowl.

The trouble with a sundial is that it doesn't work on overcast days, and it never works at night. Having taken advantage of the regular motion of the sun to make stone calendars and bowl sundials, maybe other regular motions could be tamed to help tell time. Fine sand fell pretty regularly through a tiny hole between two enclosed chambers, enabling ancient time-keepers to make an edible three-minute egg.

Egyptians used candles about 3,000 B.C., which were an improvement on the older oil lamps. Well-made candles burned at a regular rate, so markings on the candle would indicate how long the candle had burned. These made good clocks that did not depend on the motion of the sun to work. In China, a clock was devised that used a slow-burning mix of pitch and sawdust applied to a metal rod. To set the clock, you placed the rod over a brass bowl and tied a metal ball around the rod. When the sawdust mixture was ignited, it burned slowly toward the ball. Once the fire reached the string, it burned it away and the ball fell into the brass bowl. It was one of the first alarm clocks.

Dripping water has a regular motion that could be made into a clock. A simple bucket with a small hole is filled with water and suspended over an empty bucket. The water drips at a continuous rate, and when the bottom bowl is marked off, you simply peek into the bowl, note the level of water, and figure how much time has passed. Later the *clepsydra,* or water-clock, would be improved by adding systems of floats installed in the bottom bowl. As the water dripped down it raised the float, which could be attached to a marker that would then move across a clock face.

The Nile floods came regularly, and it was a serious matter to know exactly when. So directly challenged, the Egyptians had a working calendar about 2800 B.C. They had a twelve-month year 360 days long, to which they added five days at the end. This calendar has been modified several times since then, but each time the changes made were adjustments, not really drastic improvements. Still, people depend so directly on the calendar that any change was met with resistance.

Today we use the Gregorian calendar, named for Pope Gregory XIII, and it is an improvement over having to add five days to the end of every year. Still, non-Catholic people were slow to accept this new calendar. It was ordained in Rome in 1582, but caused riots all over Europe when people learned that they would go to sleep on October 4th and wake up on the 15th. People resented the church "stealing" 11 days of their lives.

The Gregorian calendar was accepted in England only in 1752, and even then there were fights in the streets. Some people just refused to change over, and for a long time both the old date and the Gregorian date were used. Today, of course, we realize that a change in the date steals nothing from us; it is only a matter of record-keeping.

Even in this century, when daylight-saving time was proposed, people reacted in an odd way. Saving-time simply shifts the clocks one hour forward in the spring and one hour backward in the fall. It was designed to save the daylight and make the normal daytime hours those hours that occurred during daytime. Without it, part of our spring mornings would be in darkness, and in the winter it would be dark long before dinner. Even so, one of our fine senators objected that with "more daylight," curtains and fabrics would fade more quickly!

Of course, with daylight-saving time certain remnants of our sundial beginnings must be shed. With apologies to Gary Cooper, I must point out that with our fancy time shifts and modern systems the sun is seldom directly overhead at high noon. For the northeastern United States, high noon coincided with 12 on our clocks only 15 days in 1993. Any regular motion can make a clock, but it has to be a motion that can be measured. It might surprise an ancient Egyptian to learn that we time our second off the regular vibrations of electrons in a rare metal. The standard second is 9,191,631,770 vibrations of electrons in a cesium beam.

Born: c. 636 BC
Died: c. 546 BC
Hometown: Miletus
School affiliation: The Ionian School
Best work: Predicted an eclipse of the sun in 585 B.C.
Best formula: A circle is bisected by a diameter
Fields of interest: Astronomy, weather, science

Thales

Quote: Know thyself

Before Socrates, Plato, or Aristotle, Thales earned the reputation as the first Greek wise man. He was a merchant from
Miletus, and founded the Ionian school of philosophy along
the western coast of what is now Turkey. Thales is often considered the father of science because he was the first to attempt an explanation of the physical world that did not rely
on gods and goddesses. Thales thought everything was made
from a basic substance; unfortunately, he thought everything
was made from water. At least he was asking the right questions.

We know nothing for certain about Thales, but stories
tell us that he founded Greek geometry and did scientific research on magnetism and static electricity. He drew conclusions from his experience, like when he observed what happens when you rub amber with wool. Once, when the olive
crop had been small for years, Thales was so confident that
he could predict the weather he bought up all the olive presses.
That year, as he had figured, there was a bumper crop, and
he became wealthy. He had made a fortune in oil—olive oil.

Thales's business affairs took him on long journeys to
great cities in Egypt and Mesopotamia, where he visited the
ancient centers of knowledge. When he returned to Miletus,
Thales shared what he knew with others. He is said to have
predicted the eclipse of 585 B.C., perhaps using astronomical knowledge he acquired from Babylonian astronomers.

Whether or not he actually made the prediction matters
less than the fact that the Greeks believed he had. People were
starting to think eclipses were caused by something other than
the sun god Apollo hiding his face. People were beginning to
reason, and the Ionian school exerted great influence.

Thales owned a salt mine and used mules to bring the
bags of salt to the market place. One little donkey always

showed up at the market looking fresh, even after the difficult trek from mine to market. The donkey's load was always much lighter than it should have been.

Thales found that the donkey was lolling in a stream that crossed the trail. Besides providing a refreshing break, the water dissolved much of the salt and lightened the donkey's load. Thales's solution, according to Aesop, was to fill the donkey's bags with sponges. When the lazy donkey lay in the stream now, his load became heavier. The donkey started skipping the dip in the stream.

Counting and Measuring

Early mathematics counted and measured. Trade and farming refined measurements of weight and volume, and the first scales appeared around 5,000 B.C. Barter and trade had caused people to compare different goods. Were two goats equal to one sheep? In Asia Minor about 650 B.C., coins were minted to facilitate trade. The basics of engineering gave rise to the science of geometry. Man's activities and needs had created the beginnings of mathematics.

The mariner's activities demanded a way to measure water depth, and better methods of star-gazing were a crucial navigation skill. Mining and smelting their way into the Bronze Age, people were forced to consider ratios of numbers. The best bronze was one part tin, nine parts copper. Astronomers needed math to predict the seasons, the floods, the feasts. Math was doing things.

Counting and measuring sprang up around the activities that were useful to human pursuits, but playing with numbers was still of no consequence by itself. People would speak of one ox, or five loaves of bread, but there was only the vaguest notion of one or five as numbers on their own. Numbers numbered things: animals, blocks to build the pyramids, the passing seasons. Soon people would turn their attention to numbers themselves, and once that happened mathematics would really have its beginning.

Great Book: The Rhind Papyrus

Just last century, archaeologists in Egypt discovered an ancient text that was written around 1650 B.C. Mysteriously titled *The Directions For Knowing All Dark Things*, it is called the *Rhind Papyrus* in honor of Henry Rhind, a Scottish scholar who bought the papyrus at a shop in Egypt in 1858. Many people refer to it as the *Ahmes Papyrus,* after the scribe whose name is on the work. This ancient scroll is actually a math text.

Written in ancient hieratic script, it is in part a catalogue of 84 math problems, many dealing with the problem of summing fractions. Ahmes probably copied it from an even older papyrus, and today this valuable find provides a glimpse of Egyptian life as well as Egyptian mathematical knowledge.

Some of the problems listed include the procedure for dividing loaves of bread among 10 men, and for figuring the amount of food required to keep a stall of oxen fat and happy for a month. These *word problems* are set out in reasoned detail, and provided examples, or recipes, that readers could use to figure out similar problems.

Really important information, like the ratio for the slope of the pyramids, is given an entire section of the papyrus. There are even references to the golden ratio, a mathematical description that may be the very basis of beauty.

Numbers

Astronomers used them every night and engineers used them everyday. Once the tax collector came on the scene, no one's life could escape the world of numbers. Even today, though you may be shed of trigonometry classes, your life is touched by numbers. What are these things, anyway?

Throughout most of human history, people have been far too busy surviving to think about it much. For most of the past in fact, slaves and ordinary people were not taught to count, much less to think about number theory. When certain numbers worked for predicting the seasons, great. When surveying techniques delivered improved irrigation, fine. When beautiful temples were erected, wonderful. Never mind what they are, they work.

We employed numbers anywhere we could but rarely thought about the essence of numbers. The concept of numbers was dim. The idea of numbers in their own right, apart from the things they counted, was coming along as slowly as meaningful scientific thought in general. But some people did think about it.

People saw order and meaning in the numbers, but still interpreted their significance in mystical, religious words. Numbers litter the Bible with special significance to the holy trinity and the seven days of creation. Special properties of numbers were likely to be explained by all sorts of superstitious beliefs, and certainly people had their lucky numbers. In about 550 B.C., a group of people gathered together on the basis of what they thought numbers were, and they thought numbers were everything.

The Pythagoreans

No one thought so much of numbers as the members of the Pythagorean Brotherhood. This gang of number-lovers gathered around the Greek mathematician Pythagoras, and it was he who said that "everything is number." Everyone at this time had lives touched in some way by numbers, but, for the Pythagoreans, numbers practically were life.

To this unusual group, numbers were male or female, wet or dry, hot or cold, right or left, light or dark. And yes, they even made distinctions between good and evil numbers. Forming a tight-knit society, the Pythagoreans lived in an ultra-conservative commune and shared all property among themselves.

Much like the American Quakers, members of the Brotherhood denied themselves most of the frills of life that were available in those days. They thought that friendship and modesty were the most important aspects of all human interrelations. They ate no meat and practiced sexual abstinence. After all, who needs sex when you think the ultimate reality of the universe is number?

Pythagoras

Born: c. 582 B.C.
Died: c. 507 B.C.
Hometown: Samos
School affiliation: The Pythagorean Brotherhood
Best work: Never published
Best formula: $A^2 + B^2 = C^2$
Fields of interest: Music, number theory
Quote: All is number

Pythagoras was born on the Greek island of Samos to a stone cutter. As a young man he taught astronomy and may have studied with Thales of Miletus. Here Pythagoras would have heard explanations of the world that did not need gods and goddesses to be part of the answer. If Pythagoras did study under Thales, he studied under the man who was practically inventing science.

After studying at the Ionian school, Pythagoras settled in the coastal city of Croton in South Italy, where he thought mostly about numbers. He thought of square numbers and triangular numbers—that is, he saw how some numbers could be arranged in triangles, and some in squares, while others could not. He computed the areas of circles and cones. Pythagoras, though, is best known for the formula that finds the hypotenuse of right triangles.

No matter that the formula was well-known before Pythagoras. His fanatic followers began a successful public-relations campaign that forever tied his name to the theorem. Pythagoras's remarkable notions about the nature of numbers attracted quite a cult following. These cults seem odd to us, but this was a time when there were all sorts of cults responding to new religious needs. Remember that Pythagoras is practically a contemporary of Buddha and Confucius.

Music was already a beautiful part of people's lives in Pythagoras's day, and the mathematician liked lute songs as much as anybody. He noticed which notes sounded best when played on a lute. Pythagoras saw that for lute strings to produce harmonious sounds, the notes played must be certain fractions of the length of the string.

The numbers seem to explain the pleasing sounds. Perhaps the numbers controlled, or ordained, the music. Numbers, it seemed to Pythagoras, controlled not only music, but everything else as well. This Pythagorean mixture of math and religion was so attractive that his following lasted many years after his death. His standing as a mathematician was undoubtedly embroidered by his strange and loyal followers.

The Pythagorean Theorem

It is definitely misleading to call it the Pythagorean Theorem. Before Pythagoras, the theorem was included in an Indian treatise on construction and temple building by Sulvasutras. Ancient Babylonian cuneiform texts also present the theorem. There aren't even any written records that would show Pythagoras worked it out on his own. Still, it was so special to the Pythagoreans that they claimed it belonged to their hero.

The familiar theorem shows the relationship between the sides of a right triangle. A right triangle is one that has a right angle, or a 90-degree angle. Ninety-degree, upright angles were important for building straight walls, so the 90-degree angle was called a right angle. It had nothing to do with politics, it's just that the 90-degree angle is the right one to use!

The squares of the short sides add up to the square of the long side, called the *hypotenuse*. The numbers 3, 4, and 5 worked for the theorem, and Pythagoras knew it. People could use this formula to figure out how many blocks it would take to build a certain-sized pyramid. Old Babylonian texts describe a problem of a ladder leaning against a wall. How far must the foot of the ladder be moved away from the wall when the top of the ladder is slid down a number of units? If you think of the ladder as the hypotenuse of a right triangle, the Pythagorean Theorem can be used to solve the problem.

This was a formula that would work for any right triangle, and could be used like a recipe, plugging in the appropriate numbers and getting an accurate answer. The form is very important, $A^2 + B^2 = C^2$ (a famous problem will come from the French mathematician Pierre Fermat in the 16th century), so the formula was more than a historical curiosity. The people of the Pythagorean Brotherhood held this formula in great reverence, but there was a problem lurking within the formula.

Their Number Was Up

To the Pythagoreans, even their eternal souls were self-moving numbers, able to pass from body to body. It is no wonder that the Brotherhood stressed the diligent study of arithmetic for all its members. For these people, the study of arithmetic was the key to perfection. Numbers could be seen as combinations of units, and the Pythagorean community searched for all the different ways numbers could be separated, combined, or compared.

It is ironic that the very theorem they so revered would rock the foundations of the Brotherhood. The theorem was true, and numbers were infallible, but the theorem also pointed to a number the Pythagoreans had never seen. It wasn't a triangular number, and it wasn't wet or dry. It was irrational, and there weren't supposed to be any of those.

The number that shook the cult was the hypotenuse of a right triangle with sides of 1 unit. By applying the theorem to this triangle, the answer for the hypotenuse is the square root of 2. What number multiplied by itself is 2? The right triangle with short sides of one unit each could be drawn, and the hypotenuse was there in the sand, but what was the square root of 2?

There is nothing to be afraid of in *irrational* numbers; they are not unbalanced. It's just that they cannot be described as the ratio of two other numbers. The decimal expansion of an irrational number will repeat forever, and today we write the square root of 2 as 1.414213562 . . ., but the Pythagoreans could not yet describe it this way.

In fact, they didn't describe it at all. They tried to hush it up, and pronounced the death penalty on any who spoke of it. Even the Pythagoreans couldn't make irrational numbers disappear, but they tried. Their whole life was wrapped up in these things, and it wouldn't do to have crazy numbers that they couldn't describe. Still, the discovery of the irrationals did not bring down the Brotherhood right away.

Rigid secrecy kept the Brotherhood alive long after Pythagoras's death, and many of their ideas remain in the way we use numbers to talk. The number four was the first perfect square, and for the Pythagoreans it stood for the Earth. You don't have to travel to the four corners of the world to find Pythagoreanism in our language.

Numbers As Shapes

Some people thought that the nature of numbers was intimately tied to shapes. A triangle has three sides and three angles; a square has four sides and four angles. Perhaps the *essence* of three is seen nowhere better than in a triangle. Of all the things that math did, building the world's great monuments was one of the most noticeable. Why couldn't the numbers be seen as eternal shapes?

According to the historian Herodotus, the yearly Nile floods had made it necessary to develop accurate surveying techniques, and from this the Egyptians developed geometry. However true this may be, by the time of the early Greek philosophers much geometry was already known. In fact, knowledge of shapes and geometry was very highly thought of by Plato. This early philosopher is said to have tacked up a sign over the door to his academy that read, "Let none ignorant of geometry enter here." When one student complained that he saw no benefit to learning geometry, Plato is said to have given him a coin, since the student had to benefit from his studies.

If the Egyptians had made geometry practical, the Greeks began to make it theoretical. Shapes in the sand were thought to represent the ideal shapes that existed outside human existence. Among the shapes that Plato studied were the five regular polyhedra, which we call Platonic solids. They are the tetrahedron, or three-sided pyramid; the hexahedron, or cube; the octahedron, which is two four-sided pyramids put together; the dodecahedron, made up of twelve pentagons; and the icosahedron, which was a monster made from 20 equilateral triangles.

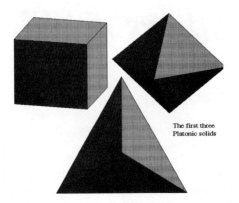

The first three
Platonic solids

These shapes were found in minerals and crystals, and were considered eternal, ideal shapes. These solids were described in detail by Plato, and were said to represent the materials that composed the universe. The four-sided cube represented earth, and the twelve-sided dodecahedron represented the universe, each face taking a sign of the zodiac.

By the time of Pythagoras, people knew a lot of things about shapes and geometry besides just the Pythagorean Theorem. In fact, geometry had been made a pretty serious science by Greeks like Eudoxus and others. For the Greeks, the concept of number was described in terms of their geometry, and around the third century B.C. their geometry was worked out and compiled by the greatest of all the Greek geometers.

Born: c. 300 BC.
Hometown: maybe Alexandria
School affiliation: Alexandria University
Best work: *Elements*
Best formula: Greatest common divisor
Fields of interest: Geometry, number theory
Euclid   **Quote:** There are an infinite number of primes

Little is known for sure about Euclid's life, but it is likely that he was born in Alexandria. It would be fitting if it were so, since he is the best-known scholar associated with that magnificent center of learning. Founded and named by Alexander the Great, Alexandria was a great seaport on the Mediterranean Sea.

Stories tell of the great library at Alexandria, and the bins that bulged with scrolls of written works. As ships docked to trade, any books on board would be taken and copied for the library. If you were lucky, you got your book back. If you were not so lucky, you might be given the copy. The library became the first great collection of the world's knowledge. Perhaps Euclid was inspired by the presence of the great scrolls, because his book *Elements* was a real bestseller.

His work was a compilation of the established science of geometry. It was such a thorough work that it was considered must reading for more than 2,000 years. *Elements* was arranged in a very orderly fashion, one idea leading to the next. Euclid also included a section that recapped the work of Thales, Pythagoras, and others.

Euclid's reasoned approach included proofs of important statements. He formalized and consolidated most of the mathematical knowledge of the day, and popularized the deductive approach. Part of the power of Euclid's work is the simplicity of its presentation.

From a few givens and unchallenged notions—like the whole is greater than the parts—Euclid added propositions and proved several hundred theorems. The theory of numbers attracted Euclid, and he proved that there is an infinite number of prime numbers. Many geometry students have cursed Euclid's name while learning geometry, but he is more famous than infamous. His work stood as a hallmark of truth, unchallenged for 2,000 years.

His formula for finding the largest number that will divide evenly into two given numbers is straightforward. Divide the larger number by the smaller and note the remainder. If there is no remainder, then the smaller number itself is the greatest common divisor. If there is a remainder, divide that into the first divisor, and keep doing this until there is no remainder. When there is no remainder, the greatest common divisor will be the last divisor.

Greatest Common Divisor

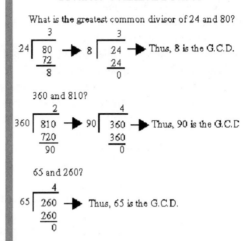

What is the greatest common divisor of 24 and 80?

Thus, 8 is the G.C.D.

360 and 810?

Thus, 90 is the G.C.D.

65 and 260?

Thus, 65 is the G.C.D.

Still, Euclid's legacy depends as much on his presentation and method as it does on the subject of geometry. Now geometry was generalized for any numbers, and geometrical theorems were presented in a reasoned way, without appeal to pre-selected numbers.

Great Book: *Elements*

The greatest of all math texts must be Euclid's *Elements*. Its 13 books begin with five essential postulates, observations that are intuitively true but not proved. From these five postulates and five stipulations (axioms), Euclid proceeds to prove almost 500 propositions about what we now call plane geometry.

Elements became famous even in Euclid's day, essential reading for all educated people. It was taught in Euclid's own words until after the American Revolution. When starting out as a lawyer, Abraham Lincoln went from town to town to eke out a living. Always looking to better himself, he spent his free time reading Euclid's *Elements*.

Elements was more than just a mathematical textbook. The spirit of *Elements* comes from Plato's high regard for mathematics, without the Pythagorean mysticism. It represented truth. It was the epitome of logical thinking.

Euclid listed outright what he assumed to be true and called them the postulates. This collection of definitions—or axioms, or common notions—was the starting block for Euclid's system. An example of an axiom would be "the whole is greater than the parts." From these few definitions, he cleverly employed strict rules to prove the rest of the theorems and postulates.

The Postulates

1. It is possible to draw a straight line from any point to any point.
2. It is possible to produce a finite straight line continuously in a straight line.
3. It is possible to describe a circle with any center and radius.
4. All right angles are equal to one another.
5. If a straight line falling on two straight lines makes the interior angles on the same side less than two right angles, the two straight lines, if produced indefinitely, will meet on that side on which the angles are less than two right angles.

That last one is a mouthful, and look how different it is from the others. Could there be something wrong with it? The book got good reviews except for this particular postulate. Its clumsy wording was a nagging problem for mathematicians that came later. That one postulate would finally tarnish the truth of *Elements,* similar to what happened when the Pythagoreans found irrational numbers. But this wouldn't happen until after the American Revolution.

Though best-known for geometry, *Elements* also had sections on conics, optics, elementary algebra, and numbers. Euclidean geometry, with some modern patchwork, is still taught as high school geometry.

Formal geometry and the study of solid shapes was a more scientific approach than the mysticism of the Pythagoreans. But the Greeks had also elevated numbers out of the rut of just doing things. Numbers were eternal truths, and part of the secret of the universe. And why not? Long before formal geometry, people were puzzled to find that some shapes were more pleasing to look at than others.

Pythagoras had discovered the relationship between numbers and music, but there was an even older connection between art and numbers. There was already a formula for producing pleasing shapes, and it was as well-known. In fact, for many years it had been incorporated into the ancient buildings and monuments that everyone saw.

But why were these shapes pleasing to early civilizations, and why don't they please us now? We don't know exactly what makes these shapes pleasing, but the answer to why they don't please us now is that they still do.

Math In Art, Math As Art

O ne hundred years ago, the German psychologist Gustav Fehner conducted an unusual experiment. He showed hundreds of people a simple card with a straight line across it. Fehner instructed the participants to draw another line perpendicular to the first, thereby dividing the line into two separate sections. The people could divide the line anywhere they liked as long as the division they made was pleasing to them. What division do you think is pleasing to look at?

You might think that Fehner would end up with as many different divisions as there are different people. Some people might think it looks pleasing to split the line exactly in half, and some might like to look at a line that has only a tiny tip sectioned off. Who is to say, after all, what's pleasing about a divided line?

Fehner performed the experiment with hundreds of people, and you may be surprised at the result of his experiment. A large percentage of the people in the experiment divided the line into very similar segments. Most people, in fact, divided the line at the place the ancients called the golden section. Did you?

Mathematically, the golden ratio is written as

$$\frac{1+\sqrt{5}}{2}$$

Put into words, a line is divided into a golden section when the smaller part relates to the larger part as the larger part relates to the whole line. This golden ratio can be found in lots of shapes other than straight lines.

Golden Shapes

Rectangles have two short sides and two long sides, and when these measurements reflect the golden ratio you have a golden rectangle. To make a golden triangle, draw an isosceles triangle (one with two equal sides); make the two equal sides the proper length to be in golden ratio to the one short side.

This ratio is often represented by the Greek letter Phi in honor of the sculptor Phidias, who used lots of golden proportions in his art. Phidias may have lent his name to it, but many before him knew the pleasing aspect of the golden ratio. Pythagoras, Plato, and Aristotle thought about the golden section, but something about this ratio struck people as pleasing long before even these old philosophers sat around discussing it. In fact, there were examples of the golden section all over the ancient world.

The pyramids are examples of golden ratio architecture, and the great pyramid at Gizeh was constructed more than 4,500 years ago. Our old friend Ahmes dedicated a couple of pages to diagrams that illustrate the proper slope of the pyramids. The height and base of the pyramid possess a ratio of about five to eight, more precision being impossible since the years have worn down

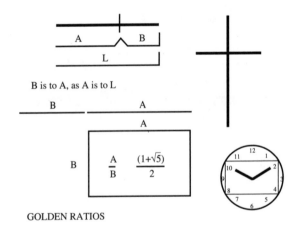

GOLDEN RATIOS

the apex considerably.

Nevertheless it is a golden ratio; each side of the pyramid could have been sliced out of a golden rectangle.

Golden buildings kept going up. About 1,400 years after the great pyramid was finished, Ramses IV passed away. Big shot Egyptians had many a fancy final resting place built for just such an occasion, and Ramses was no exception. His tomb is a nest of chambers within chambers within chambers—all built to golden proportions.

The Greeks entered the picture with the most golden of all ancient buildings, the Parthenon. It is one giant golden rectangle, and, since its dedication around 435 B.C., it has become one of the most copied buildings in the world. Later the Romans carried on the tradition (actually they stole it, the same way they got everything else), and Roman urns, vases, and statues are imbued with the golden ratio.

So mysterious was the pleasing nature of this relationship that in medieval times it was called the *divine proportion*. Renaissance painters made a conscious effort to put the golden rectangle to use. They used it in plotting the major focus of their paintings, and even in the selection of frames for the work itself.

The golden ratio can be found all through history, and, as Fehner's experiment indicated, it can be found in today's world as well. It shows up in my 3 by 5 index cards, and it can be found in the shape of ordinary playing cards and board games, the proscenium stage and wide-screen movies.

Madison Avenue knows the golden ratio. Advertisers take everything into account, and they figure that since people find these shapes so pleasing, they should try to sell stuff using the golden proportion. When people look at a clock, they will see a golden rectangle, if possible. So, in a floor wax commercial, the kitchen clock is often set at 10:10, or 8:20. With the hands in these positions, it's easy for the mind's eye to fill in the rest of the golden rectangle. (At least that's their high-priced theory.) Right or wrong, for years Dr. Pepper was recommended at the times 10, 2, and 4.

Math adds to the artist's palette in lots of ways. The use of perspective in the visual arts and the development of projective geometry borrowed from each other. Albrecht Durer helped develop projective geometry and bring it to the artwork of the day. Today the study of fractals is a critical element in the success of computer art.

Not that perspective and fractals have displaced the golden ratio. Golden shapes still appear. Tombstones and crosses can be seen as examples of the golden proportion, as well as doorways, gates, and windows. Twentieth century architect Le Corbusier was very impressed with the golden proportion. He is responsible for the design of the UN building, and we need look no further for a better example of a golden rectangle.

Why? Why are these relations so pleasing to us? For one thing, the golden ratio can be found in nature. An average chicken egg will fit snugly inside a golden rectangle. More importantly, perhaps, much of the human body can be seen to be in golden ratios.

Maybe the pleasing nature of golden proportions comes from the emphasis nature itself places on golden ratios. The human body is full of them, especially in the face. A box around the eyes and top of the mouth is an almost perfect golden rectangle. The smiling face of mother may be the first golden rectangle we see.

Math Goes To War

Ith mathematics contributed to the pleasing art and architecture of the day, it was employed in less delicate matters just as well. Mathematicians were useful to have on your side when you went to war. Even the earliest skirmishers would have benefited from having someone in the ranks who could count. Even a little math might save outnumbered troops. The math of the marketplace had to be adopted to the logistical needs of an army. If you're going to lay siege, you've got to bring along enough food and water, spears and slingshots, tents, cannonballs, and bandages.

The company clerk could handle this type of bookkeeping easily enough, but there were other uses for math skills on the battlefield. To operate a catapult, you had to know something about how a projectile will act, even if the earliest cannoneers used more knack than math. After gunpowder came to Europe about 1350, warriors began to use real cannons, and more serious mathematics helped aim them.

The modern warrior developed quite an arsenal using mathematics and science—the rifled gun barrel, mustard gas, Agent Orange, the atomic bomb. The math and science of the present century contributed heavily to the creation of new weapons for war. The horrible killing power of these new creations is just an extension of the uses generals have always found for scientific discovery. Perhaps the greatest mathematical mind of all was Archimedes. Though his devices and discoveries were often put to use in battle, he contributed to every major field of scientific inquiry.

Archimedes

Born: c. 287 B.C.
Died: c. 212 B.C.
Hometown: Syracuse, Sicily
School affiliation: Alexandria
Best work: *The Sand-Reckoner*
Best formula: Method of exhaustion
Fields of interest: Geometry, number theory, mechanics, hydraulics
Quote: Give me a place to stand on, and I will move the Earth

He was the son of a respected astronomer, and may have been kin to King Hiero II of Syracuse. Archimedes studied for a brief time at Alexandria, where he made friends with the scholar Conon of Samos. Conon had been schooled by the great Euclid himself, and no doubt Archimedes learned much from him. It is hard to estimate the importance of this great mathematician and inventor, but it is also hard to swallow some of the myths surrounding him.

When the Romans lay siege on Syracuse, it is said that Archimedes built a giant claw that could reach out and snatch a ship right out of the water. Other stories tell of a giant lens he used to set attacking ships on fire. There are grains of truth to all of these stories, but the fear and respect Roman soldiers had for Archimedes's inventions pumped up these myths.

Perhaps his most famous problem was posed to Archimedes by the King of Syracuse himself. The King had commissioned a golden crown, but thought the artisan had cheated him. How could Archimedes determine that the crown was pure gold without damaging it? Archimedes lost himself in the problem. The answer came to him as he let himself down into his afternoon bath. Archimedes noticed the water rising as it was displaced by the weight of his body.

The crown and an equal amount of pure gold could be submerged to reveal their true weights. Archimedes was so happy at his discovery that he did not bother to dress before he raced to tell the King. He ran nude through the streets of Syracuse shouted "Eureka! Eureka!" (in English, "I found it!").

Archimedes invented the water screw, which lifts water by rotation, and did experimental work with the lever. He realized the limitless power of the lever, and thought that with it and a place to stand he could move the earth. His book, *The Sand-Reckoner,* is still a trenchant introduction to the concept of the lever. He moved the Earth all right, and mathematics will never forget him.

Archimedes's contributions to the discussion of math and war are full of fantastic stories. One of his inventions was said to be able to lift a Roman ship out of the water and dump out the men and equipment. Did he do that? It is more probable that this is a story told by scared Roman soldiers. It may be based on some truth, and is not unreasonable to think that Archimedes created some kind of wench system, with block and tackle, that could lift small ships out of the water.

As far as burning the attacking ships with a giant lens, this is almost certainly untrue. Experiments to duplicate this feat have only had limited success, and that is with stationary objects, not moving ships. Once the city of Syracuse was overrun by the Romans, the soldiers were ordered to capture Archimedes alive. Unfortunately, as the story goes, the great mathematician was so involved with his drawings in the sand that he ignored the soldier's commands to surrender. Doing what soldiers do, they slew Archimedes on the spot.

Setting myths aside, the inventions of Archimedes are said to include placing small holes in city walls that defenders could safely shoot arrows from. His devices that spilled boiling oil on Romans scaling city walls were fearful things indeed. Once defenses like these were used, all the people of Syracuse had to do was drop a rope over the wall to send the Romans retreating. So great was the fear of Archimedes's weapons of war.

Other aspects of military campaigns drew on current mathematical knowledge, and these methods were not used to kill people directly. For one thing, it was important to communicate orders to your lieutenants without anyone else being able to read them. How could you do that?

Tales of Decrypt

Today's world of computerized information makes encryption methods both necessary and commonplace. Many codes we use without thinking. Electronic information carries our election results, plots military locations, and constitutes the millions of pieces of E-mail we use every day. Your ATM bank card won't work, thank goodness, unless you punch in your secret code. The scary thing is that any of this electronic information can be read and altered once you break the code.

It should not surprise you that some pretty complex math can be employed to scramble sensitive information. Electronic information may be more easily accessed than a note hidden in a secret place, but the same technology that makes these messages so available is also a powerful encryption tool. Codes are plenty old though, and computer chips are just the current form of secret messages.

People have always found a way to code things, and the first thing to do before you code something is to decide whether you want your message to be read by a very few or by as close to everybody as you can get. Road signs are a good example of a type of code that is meant to be understood by many people. Great numbers of people travel long distances these days, and most will recognize the familiar international icons that point to rest rooms, dining cars, and non-smoking areas. For the ships at sea before radio—or now, when the radio is dead—the International Flag Code consists of colorful square flags that can be run up and down a mast stay.

Coordinating the different maneuvers of many troops over a large land area will test any general. In the days before walkie-talkies, messengers were the main communication link. Smoke signals, flags, and other visual markers were helpful for battlefield movements. For detailed plans of attack, though, a detailed message was called for. With enough time on their side, the early generals would sometimes shave a slave's head and brand the message into his scalp (a little dab'll do ya). Once his hair grew out and covered the message sufficiently, he would be sent on his way. On reaching his destination, the barber would shave his head, the general would read the message, and the messenger would be killed.

Depending on how fast your messenger's hair grew, of course, the message might be useless before it was concealed by hair. To send quicker military memos, some generals branded messages into slaves' arms. The slave was told that when he reached the other camp he would be given a salve for his wounds. The bloody bandages would conceal the message until he reached the other camp, where they would remove his bandages, read the message, and kill the slave.

Thankfully for these hapless messengers, there were cipher systems that could hide a message so that even if it were intercepted it would reveal nothing. So codes are agreed-on symbols for common situations. Cattle brands are this type of code. But the real work of codes—the secret, spy type of code work—starts when you don't want everyone to know your message. You could pass secret memos, but there's always the chance of the enemy intercepting them. Of course, if your message was coded, it wouldn't matter who saw it.

Codes on the Road

American hobos had a need for such a code. Traveling around the country with all your belongings in a sack can take you into lots of unknown territory, and not everybody likes a hobo. To help out fellow hobos coming through later, coded information was left carved on tree stumps, fence posts, and railroad cars.

To anyone but a fellow vagabond, this hobo code looked like children's drawings, or simple designs and doodles. But to those who knew the code, the coded messages indicated "Danger at the next house" or "Be quiet!" The lucky hobo might come across a sign that pointed ahead to a possible handout.

Codes can be any signals or symbols that stand for something. At one time, whole books of codes were used by the military, government, and business. Of course, the best way to crack this type of prearranged code is to capture a master code book. With that, you can decipher codes as easily as the people that the message is intended for. To make this a lasting advantage however, you must keep the enemy from knowing you have their code book, for once they know they will certainly create a new code book.

Naval code books were often bound with heavy lead bindings. With your ship's defeat imminent, you simply tossed the code book overboard and sank it away from enemy hands. Many naval codes gave long service since these code books were rarely captured. Still, code books have to be changed regularly.

Ciphers

Ciphers are special codes that code the message letter by letter, substituting other letters or numbers for each letter in the message. The plain text is the message to be sent, and the cipher text is the encoded plain text. Substitution ciphers were used by Caesar to defeat the Gauls, but the earliest concealing techniques had less to do with ciphers than with the cruel disregard the military generals had for their slaves. Math to the rescue!

Julius Caesar used this kind of code. He knew the value of modular arithmetic and employed a code similar to the kind found in many daily newspaper cryptoquotes.

The Caesar Cipher was created by lining up two alphabets on top of each other and sliding the bottom one three letters to the right. The letters of the Roman communique were shifted, all coded *a*'s became *c*'s, *b*'s became *d*'s, and so on. Only a few key people knew the code's secret. The fact that a captured soldier didn't know the code would help keep the code a secret, although the lack of such knowledge didn't help him on the rack.

TRY'EM—Cipher Slides

There are lots of easy ways to make cipher slides. The most basic consists of two strips of paper. On one strip write the complete alphabet. On the other strip write two alphabets, one after the other. These letters are going to have to match up later, so graph paper makes a more precise instrument.

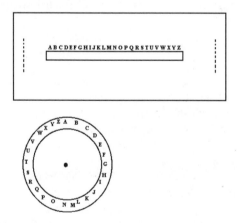

Now cut a couple of slits so the strip with two alphabets can slide against the other alphabet. When you make the horizontal slide shown here, shift the alphabet three letters to the right and you'll be using the code Caesar used to defeat the Gauls.

With the cipher slide, you can move the alphabet to the right or the left any number of spaces you like, and all will be different code systems.

The picture shows a circular cipher slide that's also easy to make, though when marking off the circular disks the sections must be very accurate so that the letters match up wherever you spin the disks. Take a look at the secret message below:

"PCYB KMPC KYRF ZMMIQ."

This message was created on the cipher slide, but the slide won't be much help deciphering it. What you need is a frequency list. Since you know the plain text message was in English, even in this coded message the letter patterns will be the same. The letter *z* is used infrequently in English, and you would expect the corresponding letter to be used just as little. There are lots of *e*'s in English, there are two *w*'s in the message. Could one of them be an *e*?

The most frequently used letters in English are as follows: E,T,A,O,N,I,S,R,H,L,D,C,U,F,P,M,W,Y,B,G,V,K,Q,X,J,Z.

With this tool, the next step is to make a frequency list of the message and start trying substitutions. If there are a lot of *q*s in the message, they may be *e*s, or *t*s. Obviously short messages are harder to decode than long ones. In long messages, there are enough letters to really see the frequencies.

Of course, different languages will have different frequency lists. The start of the frequency lists for French and Spanish are, respectively: E,A,I,S,T and E,A,O,S,N.

When you tackle the Cryptoquote in your daily newspaper, you face essentially the same problem the Gauls faced. One letter stands for another in these puzzles, and they're solved through an educated trial-and-error method. Of course, it's possible to solve each puzzle, or they wouldn't be so popular. Even though the letters are mixed, each *a* in the cipher text refers to some letter in the plain text, and if the plain text was English ....

Francois Viete made important contributions to mathematics of cubic equations, and introduced letters for variables in algebra. He made his contributions in the late 1500s, but was probably best known for breaking the Spanish code, which consisted of more than 500 ciphers. The Spanish thought their code was too complex to be broken by ordinary means, and were so astounded at Viete's accomplishment that they accused him of being in league with the devil.

Before drafting the Declaration of Independence, Thomas Jefferson tried his hand at building a code machine. Jefferson printed the entire alphabet on 36 wooden wheels, each slightly larger than the next. By spinning the dials in order, he would arrange the letters so that they spelled his message. At this point, the wheels could be locked together and turned all at one time. Any other line of letters could be used for the cipher text. When the encoded message was received, the wheels on the recipient's Jefferson machine would be made to line up to spell the cipher text. Then the wheels were spun together until the message could be read.

Jefferson's method required that both sender and recipient have the same machine. It would have been good for business if it had caught on. Pre-arranged codes were used extensively in the Civil War in America. A favorite of the graybacks was a form of cipher called pigpen cipher. The following forms were agreed on:

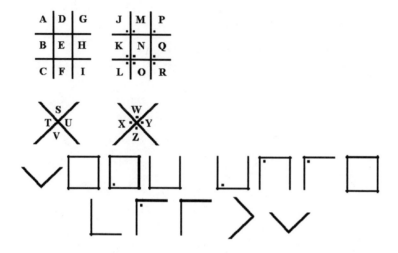

A captured Confederate message is written below the pigpen ciphers for you to practice decoding.

Death and Taxes

While you're busy decoding things, try decoding the instructions that come with this year's 1040 form. A Ry Cooder song tells us that the taxes on the farmer feed us all. Herodotus described how geometry evolved from the need of the Egyptian rulers to tax the people (they had a lot of monuments to build). The easiest, fairest way was to tax a portion of a person's land area. Land areas on the banks of the Nile change with each flood tide, though, creating a great need for surveying methods.

Using ropes with knots tied at regular intervals, Egyptian "rope stretchers" measured what was left of a farm after the floods and taxed the farmer appropriately. Of course, if the Nile flooding increased your acreage, depositing a larger area of rich soil than before, your taxes would no doubt increase appropriately. This is the way property tax began, and it is still the main source of revenue at local-government levels.

Ancient Egyptians placed a great emphasis on death, and especially the death of their rulers. In fact, most of the great monuments that survive today are tombs for or tributes to the rulers of the day. Death and taxes were obviously on their minds, and it seems that they had already made an important observation about these two events—namely that they were unavoidable. Unavoidable maybe, mathematical definitely.

The Tax Man Cometh

Relative to the situation today, taxes had little influence in the ancient world. Nevertheless, there is a long history of taxing, and basing taxes on a farmer's land area was only one way—not necessarily the most fair way. Taxing the fertility of the land is more fair.

Even taxing in kind, not currency, is more fair, with the landowner giving up, say, one-tenth of what the land produces. This one-tenth became the basis for what we know as the tithe. So, long ago, taxes were mostly consumption taxes and taxes on real estate transactions.

It worked out well, and the state got by fine on a reasonable percentage of its people's labors and holdings. A person's individual net worth and income was left alone. From time to time, the state found itself needing to spend more money—sound familiar? Big construction projects can create a deficit, but nothing raises taxes like a nice little war.

Small, everyday atrocities are easily within the budgets of most governments, but grand-scale slaughter requires a bigger war chest. There were lots of different taxes devised to raise money for war, but even then only a really big war would require raising money by imposing taxes on a person's net worth.

Until the time of Julius Caesar, taxes were collected by "tax farmers." As private businessmen, their job was to collect taxes. Like today's bill collectors, the farmers would take as their pay a percentage of the total tax collected. How popular do you think these independent tax farmers were on collection day? Caesar made tax men civil servants, which they remain today.

There were several regular head taxes—or direct taxes—in Rome, called the tributum. The base of the tributum could be expanded to include land holdings, for instance, in case the military machine required more fuel. Always concerned about family matters, the Romans also favored heavy inheritance taxes.

The Middle Ages saw the disappearance of consumption taxes and concentration on obligatory services or "aids," market fees, and transit duties. Excise taxes were important in Medieval Europe. These can take the form of taxes on goods produced. Taxes on land and houses remained strong, but only the rich ever had their income taxed. And don't forget the American Revolution. Many revolutions, in fact, were direct responses to unfair taxation.

Adam Smith's book *Wealth of Nations* was published in 1776, and it set forth four basic rules for fair and rational taxation: certain, convenient, economical, and based on an ability to pay. The British government was starting to listen to him.

Income tax was first introduced as a general rule in England in 1799. Was it to provide health care and education to the people? Was it to fund better agricultural methods or spur science and technology? Of course not. The government needed the money to finance the Napoleonic Wars.

Anyone making an income of more than 200 pounds would have to send 10 percent to the government. By the 1880s, income tax was generally accepted as necessary in England. The war machine always has its hand out, and in 1940 England introduced a purchase tax as a war measure.

Germany experimented with income tax in the 1840s, but France held out, with motions for income taxes made in government houses only around 1870. In 1909, the measure finally passed approval of the Chamber of Deputies, only to be held up in the Senate until the looming first World War forced its passage in 1914—only a few weeks before the outbreak of full-scale war.

Did you know that until 1913 April 15th was just another day in the United States? A short-lived income tax was levied by Lincoln to fuel the North's Civil War efforts, but it was dropped in 1872, after the war had been over for several years.

Grover Cleveland ran on a ticket of reducing import tariffs, and in 1894 he proposed a permanent income tax to make up the difference. That way he could uphold his political promise of reducing tariffs by reaching directly into the pockets of the voters. It was promptly ruled unconstitutional.

This forced backers to plod their way to a constitutional amendment, the 16th, which was ratified on February 25, 1913. Notice how it passed on the verge of the First World War. Until this amendment passed, customs duties had been the prime source of revenue for the government. Now, thanks to a constitutional amendment, income taxes are here to stay. As George Harrison reminds us, "If you take a walk they'll tax your feet."

Death shall have no dominion, but taxes do. There are two "popular" types of death taxes: estate taxes on the corpse's property and inheritance taxes on the inheritors' share. Throughout history, most countries have considered death a taxable event—the Romans were especially good at taxing inheritance.

There are sales, excise taxes, value-added taxes—you name it, they tax it. Of course, though income and sales tax are relatively new types of taxes, almost nothing else escaped the bony fingers of state greed. In the First Century A.D., Roman emperor Vespasian justified taxing Roman urinals with the Latin phrase "pecunia non olet"—money doesn't smell.

Strength in Numbers

Long ago, people didn't need war to remind them they lived in a dangerous place. Once ships set sail, it was a crap shoot. Many ships laden with expensive goods could be sunk by "natural" causes. They ran into foul weather, got lost in uncharted waters, or disappeared for lots of reasons. Fortunes could be lost on one trip. Wasn't there some way to hedge your investment?

Marine insurance is one of the oldest hedges in history, but all kinds of insurance have roots in the past. Annuities were available from Roman times, and by the 15th Century marine insurance was a well-established industry. Something really big would have to happen before regular casualty and life insurance was offered.

For insurance to work out for both insurer and insured, lots of information is needed. It is a relatively easy observation to see that if for every 10 ships that set out on a trading voyage about eight or nine will return, your insurance rates should be set accordingly. But how long is the normal life?

Careful, serious, statistical study of death had to wait patiently for an opportunity to study lots of deaths. The plague that swept London in the early 1600s made the study of death a terrifying necessity. We have so much statistical information today that insurance rates can be figured differently for men and women, different races, and different occupations.

Symbols

B elieve it or not, the symbols of math are supposed to make everything easier. Like codes, they are just a special type of shorthand. We could write mathematics in words—and people did before symbols were accepted—but it's a very clumsy way to represent formulas. The simple equation $2 + 2 = 4$, for example, becomes "two units added to two units is equal to four units."

Translating from words to symbols loses nothing in meaning, and helps focus on the mathematical relationship the formula describes. Even today some formulas contain partial words, like lim for limit. Sometimes symbols were adopted because they were simpler, and sometimes new symbols were needed to represent a new idea. Not just a matter of taste, most symbols gained acceptance based on their simplicity and usefulness. Outside influences like the development of movable type also came to bear on the changing set of mathematical symbols.

All Kinds of Numbers

The abacus was for calculation and symbols were used to create a permanent record. Since some of the oldest existing texts are tabulations of one kind or another, I'd say that's pretty damned permanent. The archaeological record provides lots of examples of different number systems.

Egyptian hieroglyphics from before 3,000 B.C. represent numbers by collections of unique symbols. They had a symbol for 1, and a different symbol for 10, 100, 1,000, and so on. These symbols represented a numerical value, without regard to their position. A scroll, as seen here, represented 100:

The number one was represented by a stroke:

As a result, the number 201 could be written in any of the following ways:

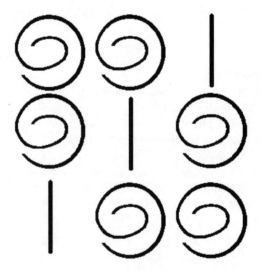

The Egyptians represented lots of numbers, some bigger than you might think they would be interested in. Big numbers must have impressed them—the glyph for 1,000,000 was a picture of an astonished man.

Roman numerals are certainly the most familiar of the ancient numbering system. When producers put the release date on their films they use the Roman numeral system. That way, a dog from 20 years ago won't be disregarded because it was made 20 years ago. It will just be disregarded because it's a dog.

The Romans marched their way of life all over the Western world, and they brought their numerals with them. They were as good as any system, it seemed, even though you couldn't really do anything with them. Flying fingers still made the abacus hum with calculations, Roman numerals entering the picture only when the final answer was to be written down. You just couldn't work with them.

Valuable Places

Even if you know that the Roman numeral XXXVI represents 36, it's impossible to derive an easy method for multiplying—or even adding—this number to another. For calculating, the Egyptian system was just as hopeless. Whatever the symbols you used for recording, you calculated on the abacus. There was a better system in use, and it was a system that you could do things with.

This system assigned different values to numbers depending on the place a number appeared. In the number 315, there are five units in the 1's place, one unit in the 10's place, and three units in the 100's place. The trouble with Roman numerals was that you had to keep adding symbols for higher numbers. With a place value system, a limited number of numerals was sufficient.

This place value notion must have occurred to abacus users. The rods on an abacus held the place 1's, 10's, 100's. When there were no beads on the bar, the bar alone held the place. Systems traveled and caught on when they were better and easier to use. Even so, the strong arm of the Roman Empire still chiseled out Roman Numerals.

Eventually the Hindu-Arabic numerals and place-value system would replace Roman numerals and other systems. The Arabs were already using something approaching modern notation, but there was another contribution to come from the Hindu by way of the Arabs. It was a great addition to mathematical symbols, yet is was small indeed. In fact, it was nothing.

*Omar
Khayyam*

Born: c. 1050
Died: c. 1123
Hometown: Nishapur (Iran)
School Affiliation: Samarkand Observatory
Best Work: *On the Difficulties of Arithmetic*
Best Formula: $x^3 + cx^2 + b^2x + a^3 = 0$
Fields of Interest: Cubic equations, algebra
Quote: The dawn is here

Ghizathudin Abulfath Ibn Ibrahim Al-Khayyâmî needed a nickname. We will call him by the name associated with his famous poetry, Omar Khayyam. *The Rubaiyat* is a collection of his poems, and it's all many remember about him. But he was also a man of mathematics.

Stories tell that as a boy Omar made a promise to his friends Nizam Al Mulk and Hassan Ibn Subbahshare. They would always share each other's fame and fortune—if any. Nizam made it big, and was true to his youthful pledge. Hassan asked to be made court chamberlain, and Nizam fulfilled the request. But Hassam did such a bad job that he was fired and banned from the court. He then became the leader of the Ismailians, who went around murdering people; "Hassam" may be the root of the word assassin. Perhaps Hassam should have been more careful what he wished for.

Omar was more careful. He asked only for a stipend large enough to provide for worldly needs. That way he could spend every moment pursuing his love for astronomy and mathematics. Nizam granted Omar's wish, and lived to see a small return on this kindness.

Omar was probably an atheist, though his writings contain many pious phrases typical of the Islamic writing of the day. His serious work is concerned with the solution of cubic equations. He greatly simplified earlier work on conic sections. Even Archimedes and Euclid had worked with conic sections, but Omar claimed to be the first to handle every type of cubic that has a positive root.

Omar was familiar with finding binomial coefficients from a mysterious triangle that we usually call Pascal's triangle. His greatest contribution was through his introduction of Arabic mathematical systems to other cultures. There would be long years before this Hindu-Arabic number system would sweep the world, but Omar gave the world its first glimpse.

The Story of 0

Nothing from nothing leaves nothing, as the pop mathematician Billy Preston has noted. Maybe that's why it was ignored for so long. It was, after all, just nothing. A place holder. Empty.

When there are no beads touching the bar on your abacus, it is because you don't have anything in that place. The rod is a place holder, and nothing more. In one of the greatest Horatio Algier stories in all math, zero began as only a place holder. It was standing and serving, but only by waiting to hold something.

It took a long time for mathematicians to use the symbol zero, but even longer for people to recognize the concept of zero as a number itself. The Babylonian astronomers were using zero, but they simply left spaces when there were no units. Zero probably traveled from the Chaldeans and Babylonians to the Hindu, from them to the Arabs, and from them, eventually, to Medieval Europeans. Far away and unknown to any of these zero-users, the Mayan civilization had been using a lozenge-shaped symbol for zero as early as 400 B.C.

Touching Bases

If we had 12 fingers, we would probably have a base 12 number system. Our 10 fingers makes the base 10 system seem natural, but other bases are actually more useful for some applications. Just what is a base anyway?

Simply put, we count to 10 before we shift to the next place in the place value system. Each symbol represents a factor of ten. Thus, in our previous example, 315 can also be seen as representing:

$$[3 \times 10^2 + 1 \times 10^1 + 5 \times 10^0]$$

The Babylonian astronomers used a base 60 system, which is much better adapted to representing the circular movement of heavenly bodies, since there are 360 degrees in a circle. For them, each symbol (actually, they used groups of symbols) represents a factor of 60. If it seems difficult, it may help to remember that you use the base 60 system every day, every hour, every minute. Telling time is a base 60 operation.

The 1s place holds the seconds, the 10s place holds the minutes, and the 100s place holds the hours. The previous example is written in base 60, like so:

$$[3 \times 60^2 + 1 \times 60^1 + 5 \times 60^0]$$

Base two is a very worthy system. It is the system used by computers, and is especially handy for them to use, since only two symbols are necessary. Since there are only two symbols, it's easy to convert symbols to physical quantities, like off and on.

Fibonacci

Born: c. 1170
Died: c. 1250
Hometown: Pisa
School Affiliation: North Africa
Best Work: *Liber Abaci*
Best Formula: $F_N = F_{N-1} + F_{N-2}$
Fields of Interest: Sequences and series, pi
Quote: There may be something to these Hindu-Arabic numbers

Leonardo da Pisa, better known as the "son of Bonacci," Fibonacci, spent his youth traveling with his merchant-tradesman father. He spent part of his boyhood in Northern Africa and was probably taught by Muslim teachers. Here he learned the Hindu-Arabic numerals, and in his later writings Fibonacci would repeatedly recommend this system over the clumsy Roman numerals.

His father's business provided him with plenty of opportunity to travel to different lands: Egypt, Syria, Greece, and Southern France. He was exposed to many different merchants and their systems for performing calculations.

Fibonacci promoted many advances in symbols, writing fractions much as we do today, though he put the fraction to the left of the whole number instead of the right. More than likely Fibonacci was copying the Arabic system, and since the Arabic language was written from right to left, the fractions end up on the left!

Picking up the thread of the great ancient mathematicians, Fibonacci was fascinated by the golden proportion. He was the first person to show the math that generates the golden proportion, and he published this formula in his book, *Liber Abaci,* around 1202. Fibonacci translated Al-Khowarizmi's book, *Hisab al-jabr wa'l muqabahah,* which introduced algebra (al-jabr) to Europe. By promoting the improved symbols he had learned as a boy, he indirectly showed that systems were translatable.

His influence continued long after he died. In many ways, his *Liber Abaci* was directly responsible for the adoption of the Hindu-Arabic numbers, though only many years after his death. So fearful were they that the new system could be used to cheat them, merchants in Florence supported legal edicts that forbade the banks to use the Hindu-Arabic numbers that Fibonacci had lauded.

Some ancient mathematical texts contain mostly words, and are that much more difficult to read. Symbols really do make it easier to work with mathematical ideas. Mathematical symbols provide a sort of time-line for the development of mathematical ideas. In the late middle ages, people were beginning to work seriously with infinite sequences. The notion of a summing of infinite operations was denoted by the first letter of the word "sum." At that time, mathematicians used the medieval long s, which looks like: f. So when you see this symbol used in a modern textbook, let it remind you of the era in which the concepts were developed.

The language of mathematics has come a long way, and now we seem to have all the symbols we need. Of course it can't last, and as new ideas come to be accepted and older concepts are reinterpreted, new and different symbols will come along. We need a lot of symbols because we have a lot of numbers. Let's take a look at the many different kinds of numbers we have.

A Numerical Walk of Fame

Certain numbers have special properties no matter what symbols you use to identify them. Pi is the ratio of circumference and diameter in all circles, and is the same every time, regardless of whether you express it as a ratio or the sum of an infinite series, or just the Greek symbol π.

Even after the numbers are stripped of all mystical associations like hot and cold and male and female, there are still some amazing properties to be found in the different kinds of numbers that we have—and we have a lot. There are odd numbers, even numbers, prime numbers, composite numbers, rational numbers, irrational numbers, negative numbers, natural numbers, whole numbers, perfect numbers, and imaginary numbers, just to list a few of the numbers mathematicians today speak about. They not only speak about them, they also calculate with them.

The following gallery takes a look at the different kinds of numbers used in mathematics. Pythagorean cult members pondered triangular numbers, but imaginary numbers are a relatively recent addition to the numerical walk of fame. A few series of numbers deserve mention, and a discussion of three individual superstar numbers will finish the section.

Cardinal, Ordinal, and Nominal Numbers
1, First, Number One

There are three types of numbers at a track meet. When you notice that 10 runners line up at the start of the marathon, you are using the cardinal numbers. Cardinal numbers are useful for answering questions like "how many?"

Once the marathon is over, the ordinal numbers are used to denote the order of the finishers—first, second, third, and so on. Ordinal numbers show up rather reluctantly in the history of language, at least in European languages.

After first, second, and third, the rest of the ordinals are, fourth, fifth, sixth, and so on. The first three ordinals are different from each other, while, from the fourth on, they follow an obvious pattern. This same pattern is obvious in other languages too, and suggests that people realized any number may be called on to represent order; thus we form larger and larger ordinals in a proscribed form.

Ordinals were used in the Olympics for sure, and maybe that's why the first three ordinals evolved before the rest. Face it: After win, place, and show, who really cares where you finished?

The identifying numbers pinned to the runners' shirts are called nominal numbers, since they are used to name things.

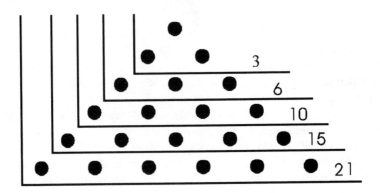

Triangular Numbers
1, 3, 6, 10, 15, ...

Back to pebbles again. These are the numbers that can be arranged in the shape of a triangle. You can see their triangular shape in the picture below. To construct the next triangular number, just add a row of pebbles onto the bottom of the preceding triangle and make it have one more pebble so it preserves the triangular shape.

This is good enough for the first few triangular numbers, but you'll need a lot of pebbles if you want to find out what the 100th triangular number is. Save your pebbles: There's a formula that's much quicker. When you want to find out the nth triangular number (15th, 16th, 100th), just use the formula:

$$\frac{N(N+1)}{2}$$

Now you can find out what the 100th or 1000th triangular number is, and you won't need tons of pebbles to do it.

Square Numbers

4, 9, 16, 25, 36, ...

These may be more familiar to you. When you square a number, you multiply it by itself. You can look at the square numbers as the series:

$2^2, 3^2, 4^2, 5^2, 6^2, ...$

The little *2* means "multiplied by itself," 2 x 2. When there is a little *3*, it means to multiply this number three times, 2 x 2 x 2.

Another interesting feature is that each square in the series can be represented as the sum of all successive odd numbers, like so:

$$4 = 1 + 3$$
$$9 = 1 + 3 + 5$$
$$16 = 1 + 3 + 5 + 7$$
$$25 = 1 + 3 + 5 + 7 + 9$$

Why not call these numbers *mm* numbers for "mirror multiplication," or something that would indicate how they're arrived at? We do, for these numbers are as square as the triangular numbers are triangular. The picture below shows the square nature of these hip numbers.

SQUARE NUMBERS

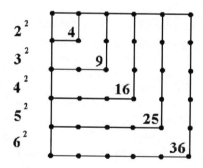

Perfect Numbers

6, 28, 496, 8128, ...

Perfect numbers, like perfect anythings, are most rare. A perfect number is equal to the sum of all its divisors, excluding itself. Members of the Pythagorean Brotherhood loved to add up the divisors of a number and compare the result to the original number. The first perfect number has the following factors (divisors): 1, 2, 3. The number itself is excluded, so the factors of six, excluding six, add up to six. Perfection!

It's easy to find numbers that are "imperfect," but can you figure out what the fifth perfect number is? The Pythagoreans couldn't. They did find numbers that were abundant, deficient, and amicable. When the sum of the factors of a number is less than that number, that number is said to be deficient (8 is more than $1 + 2 + 4$). On the other hand, an abundant number is one that is less than the sum of its factors (20 is less than $1 + 2 + 4 + 5 + 10$).

When the sum of the factors of two different numbers equal each other, they are said to be amicable. The first two amicable numbers are 220 and 284. Feel free to work out the factors on your own.

The four perfect numbers listed above were the only ones known before the 15th century, and no wonder. The fifth perfect number is a hefty 33,550,336. In 1979, the 27th perfect number was found to contain 13,395 digits!

Prime Numbers
2, 3, 5, 7, 11, 13, 17, ...

Some numbers couldn't be put into the shape of a square or triangle, and they were far from perfect. Since the only factors they have are themselves and one, the sum of the factors of any prime number, excluding itself, will always be one.

Non-prime numbers are called composite, since they're "made up" from the primes. Each composite number has a unique set of prime factors. The number *8* has the prime factorization 2 x 2 x 2, and no other number has this same factorization. Every composite number can be expressed as the product of a unique combination of prime numbers. It's this property that allows us to consider the primes as the building blocks of the other numbers, much like we consider the elements to be the building blocks of all matter.

Primes are so important that no branch of mathematics can proceed without them. For this reason, they have been the focus of intense mathematical work from the earliest times. Cambridge mathematician G.H.Hardy said, "I cannot imagine anyone being more interested in anything than I am in the theory of primes." How do they behave? Where are they? How many are there? Thanks to Euclid, we have an answer to the last question.

Great Theorem: The Infinitude of Primes

Euclid proved that there is an infinite number of primes, and his proof is straight-forward and strong. To begin with, he points out that either there is an infinite number of primes or else there is a finite number of primes. Euclid figured that if he assumed that there *was* a finite number of primes, and he could show that this assumption leads to a contradiction, then the assumption must be wrong. Since primes either are or are not infinite, and if assuming there is not an infinite number of primes leads to a contradiction, then they must be infinite. This form of argument is known as *Reductio ad Absurdum*—reducing to absurdity.

If there isn't an infinite number of primes, then it's possible to construct a list of all primes. We can pretend the list looks like so: A, B, C, ..., D. A is the first prime, B the second, and so on until you reach the last prime, D. The three dots represent all the primes between C and D. Since this list is assumed to contain all the primes, we only have to find a prime that isn't on the list. If we can do that, we will have our contradiction and therefore prove that the primes are infinite.

To begin with, Euclid created a number that wasn't on the list. He took all the primes on the list and multiplied them all together, like so: A x B x C x ... x D. To the product of all the primes, Euclid added 1 and called this number N, so that:

$$(A \times B \times C \times ... \ D) + 1 = N$$

Now we have a number that is not on the list of primes. It's clearly bigger than the biggest prime, D, since we multiplied by D and D was supposed to be the last and biggest prime. Is the number N prime or composite?

Since all numbers are either prime or composite, N must be, too. If it's prime, then there is a prime bigger than D, and, since we assumed that D was the last of a finite list of primes, there must be something wrong with the assumption. Therefore the primes are infinite.

What if N isn't prime? If not prime, N must be composite, and composite numbers have at least one prime factor. Since we multiplied all the primes together and then added *1*, none of the primes on the list are factors of N. When N is divided by A, the *1* that was added will show up as a remainder. In fact, if you divide by any number on the list, there will be a remainder of *1*. But to be composite, N must have prime factors, and we assumed that all the primes were on the list A, B, C, ..., D. If N is composite, its prime factor must lie between D and N. So, if N is composite, there must be a prime not on the list of all primes, and that is an absurdity.

The power of the proof lies in the fact that the new number, N—created off the list that was supposed to include all primes—leads us to absurdity, whether *N* is prime or composite. Therefore there must be something wrong with our original assumption, which was that there is a finite number of primes. It's a proof that a finite list of primes is impossible.

Since they are so important, it's nice to know we have an infinite number of them. Still, where are they? How can we find the prime numbers? It's easy to discover whether a number is even or not: Just divide by 2 and see if there is a remainder. If there is a remainder, the number isn't even. Finding all the factors of a number isn't so easy. To test whether the number 117 is prime or not, you must discover whether any other number can divide into 117. You could start dividing by asking, 'Does 3 go into it?,' then 4, and so on. Isn't there an easier way?

TRY 'EM—The Sieve of Eratosthenes

Given any number, there's a sure-fire way to sift out the primes up to that number. It's called the Sieve of Eratosthenes. Searching for primes among the numbers requires the proper tool. Alexandria's talented first librarian, Eratosthenes, developed a foolproof test for finding primes. It's a sieve that sifts the composites out of a list of numbers and leaves only the primes.

The sieve has an upper limit in that you must pick a number first and then sift out all the primes between 1 and that number. The sieve won't find primes on a general basis, but only within the boundary you set. Try using Eratosthenes's sieve to locate all the primes from 1 to 100.

Make a list of all the numbers from 1 to 100. A prime is a number that isn't 1, and has no factors except 1 and itself. So cross out the 1, because it's not prime.

The next number is 2, and it's obviously prime. Since there are no other numbers before it, there can be no factors for it. The 2 can be circled as a prime. Now any number that is a multiple of 2 will not be prime, so cross out all even numbers (all the even numbers are divisible by 2).

The only number before 3 is 2, and it's not a factor of 3, so circle 3 as a prime, and cross out all multiples of 3. Now back to the beginning. The next number is 5. Can you finish the sieve?

1	2	3	4	5	6	7	8	9	10	11	12	13	14	15	16
17	18	19	20	21	22	23	24	25	26	27	28	29	30		
31	32	33	34	35	36	37	38	39	40	41	42	43	44		
45	46	47	48	49	50	51	52	53	54	55	56	57	58		
59	60	61	62	63	64	65	66	67	68	69	70	71	72		
73	74	75	76	77	78	79	80	81	82	83	84	85	86		
87	88	89	90	91	92	93	94	95	96	97	98	99	100		

Now, you may want to generate more primes than occur in the range 1 to 100. The sieve works just as well, though using it to find the primes among the first 1000 digits is a time-consuming task.

Prime Time

It'd be nice if there was a formula for generating primes, but there isn't. Lots of candidates have been proposed in the past, but they're all flawed. Mersenne primes, named for Friar Marin Mersenne who found them, are of the form:

$2^P - 1$ (where p is any prime number)

But some numbers generated by this formula aren't primes.

Other questions catch our attention. There are lots of twin primes, like 3 and 5, 11 and 13, 10,006,427 and 10,006,429. Anytime there's only one even number separating primes, they're called twin primes. Is there an infinite number of twin primes? Many people think so, but no proof of this has been found. It's not even proven yet that the twin primes show up at regular intervals.

Trying to pin the primes down is frustrating. Lots of patterns appear, only to evaporate later. Consider the following numbers: 31, 331, 3331, 33331, 333331, 3333331. All are primes with a clear pattern—are all numbers of this form prime? The next one, 33333331 is prime. Unfortunately, the next number, 333333331, is not. It has a prime factorization of:

17 x 19,607,843

Prime-hunters have had more success searching for the distribution of primes. Where, in general, are the primes? From the sieve of Eratosthenes, we know that there are four primes in the first 10 numbers, 25 primes in the first 100 numbers. In the first 1000 numbers, there are 168 primes. The following chart shows the distribution of primes more clearly:

PRIME NUMBERS

Numbers (N)	Number of Primes up to N	Percentage of Primes in N
10	4	2.5
100	25	4.0
1,000	168	6.0
10,000	1,229	8.1
100,000	9,592	10.4
1,000,000	78,498	12.7
10,000,000	664,579	15.0

The percentage of primes increases by approximately 2.3

Since 2.3 is close to e (2.718), the prime number theorem says the percentage of primes in N is approximately :

$$\frac{N}{\log N}$$

Each power of 10 shows an increase in the number of primes of about *2.3*. The great mathematician Gauss thought about the distribution of primes and arrived at this conclusion when he was only about 15 years old. Approximately 100 years later, two individuals provided a rigorous proof of what is now know as the Prime Number Theorem.

Natural Numbers

N = 1, 2, 3, ...

What could be more natural? One two, buckle my shoe. The first numbers we learn are the natural numbers, because these are the first numbers we need. This simple, infinite list contains the numbers we use to count separate things. Zero isn't usually included in the natural numbers, but not because we don't use zero to count with. How many times did you kick your dog this morning? Often mathematicians want to talk about the counting numbers excluding zero, so we leave it off the list of natural numbers.

Integers

z = {..., -2, -1, 0, 1, 2, ...}

As you can see, the natural numbers are a part of the integers. The integers are the natural numbers plus zero and the negative numbers. Zero is the origin point for both infinite series of positive and negative numbers.

Negative numbers were only grudgingly admitted to the world of math. They are the additive inverses of the positive numbers, which simply means that if *a* is a positive number, and *a* + *b* = *0,* then *b* is the additive inverse of *a.* If *a* is 3, then *b* is -3. Every positive whole number has a negative counterpart that, when added to the positive number, will equal zero.

It was only about 700 A.D. that the Hindus introduced the negative numbers. They were needed for—you guessed it—representing monetary debts. Mathematicians also had a use for the negative numbers, and would use them when they had to, but until very recently the negatives did not enjoy full status as numbers. In the 1500s, Cardano used negative numbers, but he called them "numeri ficti." Even as "modern" a mathematician as Rene Descartes considered the negative numbers false numbers.

Today the negative numbers take their place as a major part of the integers. Perhaps they would have been more important to earlier mathematicians if those mathematicians had run small businesses.

Rational Numbers

Q = {..., -111, ..., -7/2, ..., 0, ..., 1, ..., 17/3, ..., 163, ...}

The rational numbers include the integers and some of the fractions. Though their name may suggest level-headed numbers, the rational aspect of these numbers comes from the fact that it's possible to represent any of these numbers as ratios of whole numbers.

Some proper fractions are 1/2, 5/8, and 1/3. They are proper because the numerator (the top number) is smaller than the denominator (the bottom number). Improper fractions—sometimes unkindly called vulgar fractions—have a numerator larger than their denominator, like 2/1, 8/5, and 3/1. The big numerator allows these fractions to be reduced to 2, 1.6, and 3.

Any fraction, proper or not, is a rational number if it has an infinitely repeating decimal. When you perform the division indicated by the examples, 1/2, 5/8, and 1/3, you get the following decimal numbers, .5, .625, and .33333 ...

The last decimal can be calculated to any degree of accuracy you wish, .3, .333, or .33333333. And, no matter how far you take the division, you'll never add any number but 3. Even the decimal number .5 can be seen to be a repeating decimal by adding zeros, .50, .5000, .500000, and again you can carry this business as far as you like.

Any decimal expansion of a rational number will be seen to repeat, even if it's only more and more zeros. A rational number like 5/7 repeats a six-digit number, .714285714285714285 ...

So the rational numbers include the integers and all repeating fractions.

Irrational Numbers

√2, √3, Π, ...

There's no reason to panic over the irrational numbers, and you don't have to keep your eye on them. They are just numbers that can't be represented by the ratio of whole numbers. Irrational numbers have decimal expansions that don't repeat.

How far do you have to take the division before you can say that nothing repeats? After all, who's to say that, after you've computed the square root of 3 to 5 million digits, those 5 million digits will repeat? Deductive proofs can prove the irrationality of a number.

There are two distinct groups of irrational numbers: the algebraic and the transcendental. The square root of any number other than a perfect square is an algebraic irrational. The square roots of perfect squares (4, 9, 16, ...) are rational numbers, but the roots of every other number (5, 6, 7, 8, 10, 11, ...) are irrational numbers. They are decimal expansions that never end and never repeat.

Algebraic irrationals answer a question of algebra. The decimal expansion 1.414214 is the rational approximation used to answer the algebraic question What is the square root of 2?

Other irrationals don't answer algebraic questions, like pi, and e. These two transcendentals are important enough to have entries of their own later. Transcendental irrational numbers have decimal expansions that don't end or repeat, and these numbers aren't roots of any other numbers.

Real Numbers

R = All the rational and irrational numbers

Every type of number so far is part of the set of real numbers. The whole numbers and the natural numbers are part of the integers, and the integers themselves are a part of the rational numbers. Even the irrationals have their place within the real numbers, and the whole set of real numbers can be seen to lie on a number line, like so:

A REAL NUMBER LINE

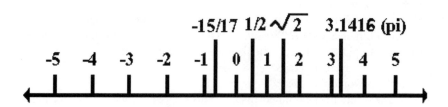

A diagram of the relationships between the real numbers presents them as a ries of sets nested within sets, like so:

THE REAL NUMBERS

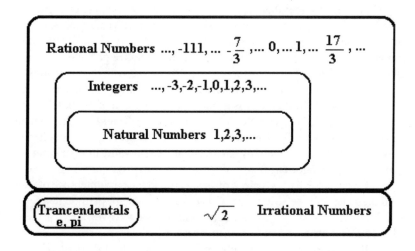

So that's all the numbers, right? After all, what would an "unreal" number look like? Would they not be "imaginary" numbers?

Three Superstars

It would be hard for mathematicians to do without any of the numbers, but the following three numbers have such important properties that they assert themselves as the superstars of the numbers. Their importance can't be overstated, and just when you figure them out, they pop up where you didn't expect them. These three superstars are pi, e, and i.

Pi

3.141592654

Would you be surprised if your television program tonight was interrupted for a special bulletin about pi? "We interrupt this program for a special news break. The number pi has just been calculated to more than 2 billion decimal places. Stay tuned for further details."

What would that mean?

The Chudnovsky brothers of New York City have indeed taken the number pi to a remarkable, world-record 2.16 billion decimal places. It took an apartment full of computer equipment and the indomitable spirit of the Chudnovskys to do it, but it's done. Why?

Pi is a transcendental irrational, which makes it sound like a guru with a shotgun just walked into a McDonalds. As a member of the irrational numbers, the decimal expansion of pi doesn't repeat, and it's transcendental because pi isn't the root of a finite algebraic equation.

Pi is the number that defines the ratio between the diameter and the circumference of any circle, no matter what size circle you care to look at. In fact, it's the essential ingredient in circles, and certainly one of the oldest "special" numbers. Potters were probably among the first to notice pi, but once the search for an exact value of pi began, it wouldn't end, even with the Chudnovskys.

You might try to find pi by measuring, as the earliest pi hunters must have done. Measuring the diameter is a cinch, but laying a ruler around the circumference of a circle is inaccurate. A description of the Temple of Solomon in the Old Testament gives 3.00 as the figure for pi (1 Kings 7:23—the molten sea is a bowl!). Though considerably smaller than the real value of pi, it's not too far off for pre-calculator days.

If you carefully lay a piece of string around the circumference of a circle, and then measure the string, you will arrive at a value of pi, though a few tries might convince you just how inaccurate this method is. If you arrive at the value 3.141592654, you are a very careful measurer!

By contrast, it's a relatively simple matter to find the area of triangles and rectangles. With a compass and straightedge you could build squares out of these figures—squares that would have the same area as the shape you started with. Around 440 B.C., Hippocrates squared the lune, which is a portion of a spherical area, like a quarter-moon. If only you could construct a square that had the same area as a circle, then pi would be easily calculated—without the crudities involved with direct measurement.

It would take a great mind to figure a way to "square the circle," and so comes the indomitable Archimedes and his method of exhaustion. Archimedes thought he could get closer and closer to the value of pi by overlaying a circle with polygons. The polygons would yield their areas easily, and, if enough were laid outside the circle and enough were laid inside the circle, pi would be trapped in between.

Using an incredible 96-sided polygon, he placed the value of pi somewhere between 22/7 and 223/71. Converting these fractions to decimals, Archimedes's method placed the value of pi like so; 3.142857 > pi > 3.14084. The polygons that enclosed the circle give the upper limit, and the polygons inside the circle give the lower limit.

A little less than 400 years later, Ptolemy arrived at the closer value 377/120. Ptolemy's method, in essence, used a 360-sided polygon. Others made attempts through the years, and Fibonacci contributed to the cause in 1220 with his figure 3.141818.

In 1579, Francois Viete found the correct value of pi to nine decimal places using a polygon with 393,216 sides! Viete's calculations, using the methods of the 1500s, were long and tedious hand calculations. What was encouraging about Viete's attempt was the method he used. It seemed that squaring the circle had reached the upper limits of calculation—a 393,216-sided polygon, for goodness sake!

Viete came up with a formula that generated pi instead of calculating the areas of thousands of polygons that approximated pi. Viete also introduced the symbols for unknown variables, x, y, and z. Though others would continue to use Archimedes's method of inscribing polygons inside and outside a circle, many more would search for better, quicker series. Viete's formula is presented below, and you can see that it's the product of an infinite multiplication. There were lots of different series that generated pi to come in the future, and Viete's was accurate to nine decimal places, though it's a complicated nest of square roots.

$$\frac{2}{\Pi} = \sqrt{\frac{1}{2}} \; \sqrt{\frac{1}{2} + \frac{1}{2}\sqrt{\frac{1}{2}}} \; \times \; \sqrt{\frac{1}{2} + \frac{1}{2}\sqrt{\frac{1}{2} + \frac{1}{2}\sqrt{\frac{1}{2}}}}$$

Isaac Newton was certainly not going to be left out of the hunt for better pi, and he published a value correct to 15 decimals in the 1600s, though this figure wasn't published until well after his death. Newton made some sheepish remarks about how the task had consumed him, and there's a good possibility that he had tried to take the figure much farther than the figure he published.

About the same time, another Englishman, William Oughtred, began to use the familiar Greek symbol for the circle's essential relationship. Euler also used the symbol for pi, and this great mathematician's reputation lent considerable weight to the symbol being accepted by everyone. In 1748 Euler published several series that converged to pi.

In 1705, 102 years after Viete, Abraham Sharp took pi to 72 decimals. The next year it was taken to 100 digits, and, in 1719, De Lagny calculated this famous ratio to 127 places. Finally, in 1766, the German Johann Lambert proved that pi was irrational. This means that you'll never find a fraction that will exactly produce pi. Still the search continued; irrational or not, the pi hunters would not be put off the scent.

A few years later, the hunters found 140 places. In 1844, they found 200 digits, 11 years later 500 digits. A high school teacher, William Shanks, culminated 15 years of calculations by publishing pi to 707 decimal places.

Irrational pi was proved to be a transcendental number in 1882 by Lindemann. Effectively ending the quest to "square the circle," the proof of the transcendence of pi made it clear as well that pi was not the answer to an algebraic formula. Did that stop people from computing pi?

In the 1940s, Ferguson found that Shanks had made an error in the 527th decimal of pi, causing every following decimal to be wrong. If you think it was crazy for Shanks to hand-calculate pi for 15 years of his life, what can be said about the man who checked his work? Ferguson published a corrected pi to 606 places, and used a simple desk calculator to take it finally to 808 places.

In 1949, the task was taken over by computer when the ENIAC found more than 2,000 decimals. Then programmers tried to find pi with each better computer, taking the challenge to 3,089 decimals in 1955, 7,480 in 1957, 16,167 in 1959, 100,000 in 1961, 250,000 in 1966, and 500,000 in 1967.

The search for the largest correct value of pi intensified with the proliferation of the computer, and records were held for shorter and shorter times. And David and Gregory Chudnovsky, as of the printing of this book, reached 2.16 billion decimal places of pi. Why?

Why is an especially good question when you consider the fact that only a few decimal places are sufficient to build anything you can think of on Earth. Some scientific work might require a dozen or so decimals. As the science writer Isaac Asimov figured, if a sphere 10 billion miles across were constructed using the early figure 355/113, any error would be only on the order of about 3,000 miles. If the 1617 figure with 35 decimal places were used, the error would be no more than the diameter of a proton.

So why pi? Well, it's an important number, and not only for the crucial role it plays with circles. Pi plays a role in many branches of analysis, and, as we will see later, pi turns up in probability theory. Why are there so many infinite series that converge toward pi? Are all of the numerals 0-9 used equally in pi, or does the number 5 show up more often, or 7? If pi is so truly transcendent, aren't its digits random numbers? Is there any pattern at all to pi?

Obviously we will improve our chances of answering questions like this if we have lots of digits for pi. Side benefits abound, and especially in the case of the Chudnovskys. You might think that it's as simple as setting a computer to the task, sitting back and watching pi appear, but it's not. Computer time is precious, but the Chudnovskys produced their figure using much weaker computers than are in existence today.

The Chudnovskys' program, which borrows some ideas from Ramanujun, converges quickly to pi, cutting down on computer time. Another important aspect of the Chudnovskys' effort is that they employ a "compute-but-verify" system, which checks the number as it produces it. Furthermore, the algorithm expresses pi as a sum, and is expandable in the sense that if you want more digits you can add them without having to start the program all over again.

The computer reigns today in the search for large values of pi, but creativity and persistence—at least in the case of the Chudnovskys—can be more important than powerful computers. The Egyptian scribe Ahmes recorded a fairly accurate expression of pi in the Rhind papyrus, but the Chudnovskys didn't record their figure on scrolls. It's on computer disks.

e

2.7182818...

This is a pretty curious number and, depending on how much mathematics you know, you may not have met e before. Like pi, it's a transcendental irrational, inexpressible by any finite combination of integers. Sometimes called Euler's number, it comes to bear in logarithms, compound interest, and number theory.

Listed below are several formulas that approximate e:

$$e = \lim_{n \to \infty} (1 + \tfrac{1}{n})^n$$

$$e = \frac{1^0}{0!} + \frac{1^1}{1!} + \frac{1^2}{2!} + \dots$$

$$e^x = \frac{x^0}{0!} + \frac{x^1}{1!} + \frac{x^2}{2!} + \dots$$

If you remember the prime number theorem, you notice that e is very close to the distribution of primes.

It's that number whose hyperbolic function is 1. A hyperbolic function is to hyperbolas what a trigonometric function is to circles. It's the base for the natural logarithms.

imaginary numbers

I

Imagine there's no answer. It isn't hard to do. The Pythagoreans screwed their foreheads up over the square root of 2. They knew it was there in the hypotenuse of a right triangle with unit sides, but that wasn't enough to make them embrace it. It seemed wrong somehow.

The Brotherhood would have really flipped over the square root of -1. It's an imaginary number. If you recall how signed numbers (positive and negative) are multiplied, you'll remember that the only way to multiply so that the product is a negative number is to start out with one positive and one negative number. Two negative numbers multiplied yields a positive number, and two positive numbers multiplied produce a positive number. How can there be a number which, when multiplied by itself, yields a negative number?

Real numbers alone are not enough to answer all the questions of algebra—$x^2 + 2 = 0$, for instance. The equation is satisfied only when x^2 is equal to -2, and, once again, how can a number multiplied by itself produce a negative number? It may seem hard to swallow—and it was for many mathematicians—but the things were useful. And that, in the end, was enough.

In the 16th century, Cardano of Italy showed how imaginary numbers could be used to solve problems, yet Rene Descartes named them "imaginary." There's a way to view all the numbers as imaginary!

Numbers that are part real, part imaginary are called complex numbers, and written like so: a + bi. The imaginary part is the *i;* everything else is real. If you like, you can attach an imaginary part to any number, as long as the imaginary part equals zero. The Swiss mathematician Jean Robert Argand graphed the real and imaginary numbers together in what is sometimes referred to as an Argand diagram. Finally, the great Gauss showed that when the real numbers, imaginary numbers, and part-real/part-imaginary numbers were taken together, all polynomial equations would have solutions.

Gauss called this big set of numbers the complex numbers, and the solutions it promised soon turned up in all sorts of practical applications—imaginary or not! Projecting 3-D information onto a two-dimensional plane can be accomplished with complex numbers. William Rowan Hamilton, the Irish mathematician, had used complex numbers to map rotational changes around a fixed point. The 18th-century interest in electricity took his notion of complex numbers as a tool to describe alternating current.

Now we can add the complex and imaginary numbers to our diagram of the real numbers, like so:

THE COMPLEX NUMBERS

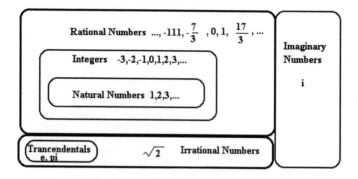

Rational Numbers ..., -111, $-\frac{7}{3}$, 0, 1, $\frac{17}{3}$, ...

Integers   -3,-2,-1,0,1,2,3,...

Natural Numbers 1,2,3,...

Trancendentals
e, pi

$\sqrt{2}$ Irrational Numbers

Imaginary
Numbers

i

Born: c. 1707
Died: c. 1783
Hometown: Basel, Switzerland
School Affiliation: St. Petersburg Academy
Best Work: *Mechanica*
Best Formula: $e^{\pi i} + 1 = 0$
Fields of Interest: Nothing much escaped him
Quote: Sir, $A + B^N/N = X$, hence God exists!

*Leonhard
Euler*

The most prolific mathematician in history, Leonhard Euler
was born April 15, 1707, in Basel, Switzerland. He would later
swamp the world of publishing by releasing 866 books and
articles between 1726 and 1800. Euler's work accounted for
almost one-third of the scientific and mathematical literature
published during that time.

At age 19, Euler won a prize offered by the French
Academy for solving a nautical question. Euler presented an
analysis of the optimum placement of masts of ships. Impressive as this task was for a young genius, it was the more incredible since Euler won the prize without ever having seen
an ocean-going ship.

This talent for numbers was obvious to his teacher,
another great mathematician, Johann Bernoulli. It's a good
thing, too, for Euler's father was determined to see young
Euler follow in his ministerial footsteps. Thankfully, Bernoulli
pleaded with the elder Euler to let the boy's great mathematical gifts flourish. His father consented, and another mind was

steered into the open seas of knowledge.

Euler's book *Mechanica* was published in 1736; it took Newton's work on dynamics and presented it in mathematical form for the first time. As great as Newton's ideas were, Euler made them mathematically workable. So productive was Euler he misplaced many of his manuscripts, and they were found only after his death.

When he was working on mathematics, he would quickly scribble his thoughts and put the paper on top of an ever-growing stack. If asked for a paper to publish, he would take one off the top of the stack, forever confusing the order of his work.

His use of mathematical symbols lent his considerable reputation to them and made them acceptable. It was as if, since Euler used it, it must be worthwhile. Among Euler's work are seen the symbols for pi, i, and summation notation.

He pioneered the field of topology, and turned number theory into a respectable science. He spent the last years of his life in Russia, and continued to contribute even after he went totally blind.

Euler's formula

As if he hadn't done enough already, Euler gave us one of the greatest formulas of all. It expresses the interrelationship between the three superstars of the numbers—pi, e, and i—all in one formula, as well as *1* and *0*. Euler created a real one-stop formula.

$$e^{\pi i} + 1 = 0$$

The American mathematician Benjamin Pierce addressed his Harvard students about Euler's formula saying, "Gentlemen ... we cannot understand it, and we don't know what it means, but we have proved it, and therefore, we know it must be the truth."

Numbers in Series

There are lots of series of numbers, and mathematicians have named them. An arithmetic series is a series of numbers that preserves the difference between each number in the sequence. You can create an arithmetical series by just adding three each time, like so: 1 + 4 + 7 + 10 + 13 ...

A geometrical series is one in which the ratio of one number to the next is preserved. You can create a geometrical series by repeatedly multiplying by 3 like so: 1 + 3 + 9 + 27 ...

A power series is one in which all the terms are regularly increasing powers. You can create a power series by repeatedly squaring numbers like so: 1 + 4 + 9 + 16 ...

An ancient Indian legend relates the power of series, in this case a geometric series. King Shirham wanted to reward Sissa Ben Dahir for inventing the game of chess. Sissa Ben Dahir used the chess board to define his reward, asking the king to place one grain of wheat on the first square, two grains on the second square, four on the third, and so on. The king was flabbergasted. Surely his subject could think of a more suitable gift for so great a creation as chess.

Presenting his loyal subject with his reward proved just how big a gift Ben Dahir had asked for. On the ninth square, the king had to place 256 grains of wheat. On the 17th, he would have to have placed 65,536 grains. On the 21st square there would have been more than a million grains of wheat. By the 41st square, Ben Dahir would get more than 1 trillion grains of wheat. There are 64 squares on a chess board, and on the entire chess board the king would have to place 18,446,744,073,709,551,615 grains of wheat.

Each series appears infinite, whether geometrical, arithmetic, or power. It may look as if, eventually, all series will end up at the same place, namely infinity. Think again. The following sequence heads toward a much smaller number. Do you know what it is?

1 + 1/2 + 1/4 + 1/8 + 1/16 ...

Fibonacci numbers
1, 1, 2, 3, 5, 8, 13, 21, ...
The above numbers are more than 325 million years old. Say what?

Ratios of the Fibonacci numbers can indeed be found in the fossilized shell of the Paleozoic chambered nautilus. These ancient mariners, as well as their modern descendants, exhibit Fibonacci numbers in the delicate spiral of their curved shells.

Plenty of Fibonacci numbers can be found in nature—by counting the numbers of seeds on the head of a sunflower, the legs of a starfish, or the petals of a daisy. Even the keys of a piano lay next to each other in Fibonacci ratios. What is this strange series of numbers?

The Italian merchant man Fibonacci developed these numbers to figure the future of a warren of rabbits, at least in part. This series of numbers was certainly known before Fibonacci, but through his popularization of them in his 1520 book *Liber Abaci* they have retained his name.

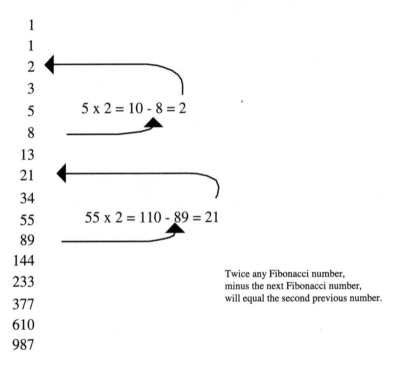

1
1
2
3
5
8
13
21
34
55
89
144
233
377
610
987

$5 \times 2 = 10 - 8 = 2$

$55 \times 2 = 110 - 89 = 21$

Twice any Fibonacci number, minus the next Fibonacci number, will equal the second previous number.

Rabbits produce one new pair of rabbits at the end of their second month of life, and another pair every month after that. If you start with newborn rabbits in January and no rabbits die, how many rabbits can you serve at the next Christmas dinner?

Actually, this information isn't intended for breeding purposes, since rabbits don't produce this way, and there's no real indication that Fibonacci ate rabbits. Still, this is essentially the problem Fibonacci tackled.

You'll only have the one pair until the end of February, then two pair in March. Your new pair won't have babies for a whole month, but the original pair will, so in April you'll have three pair of rabbits. You'll need room for five pair of rabbits in May, and eight in June—and by Christmas you can invite as many friends as 144 rabbits will feed.

If you get so attached to them that you start to call them bunnies, you might not be able to kill, skin, and cook them. In that case, you could wait until Easter, when you would have 987 bunnies that you could sell to Christians.

Fibonacci numbers are much easier to generate with pencil and paper than with rabbits. The series starts with 1. Each successive term in the series is the sum of the two previous terms, and, since we start with only one term, the second term is also 1. The third term is now 1 + 1, which gives 2. Now add the last two terms to get the next term, producing the series, 1, 1, 2, 3, 5, 8, 13, ...

Da Vinci and Kepler both knew the Fibonacci numbers and the formula that generates them, $F_n = F_{n-1} + F_{n-2}$. When you want to find the 10th Fibonacci number, just add the eighth and ninth together.

Around 1830, a man named Braun commented on the number of ridges on a pine cone. He noticed that they grew in Fibonacci ratios. A Fibonacci ratio is any two Fibonacci numbers taken together: 8/5, 13/21, 5/13. In 1920, Oxford biologist A.H. Church recorded his observations of these ratios showing up in the spiraling seeds on the head of a sunflower.

In fact, many plants grow their leaves in Fibonacci ratios. Generally speaking, leaves don't grow directly above one another; otherwise they'd block the sunlight needed by the lower leaf. Find a branch of a plant and count the leaves to see if it has a Fibonacci ratio. Count how many leaves there are, and how many times you spiral around the branch before another leaf grows exactly above the leaf you started with.

If you counted eight leaves in three spirals, you've counted a Fibonacci ratio. The leaves of an elm tree grow with a 2/3 Fibonacci ratio, 1/3 for the beech tree. Petals sometimes demonstrate these ratios: the lily with 3 leaves, corn marigold 13, field daisies 34. Even branches shoot off in Fibonacci ratios, but more than phylotactic vegetation reveals the numbers of this curious series.

Ebony and ivory occur in Fibonacci ratios, at least on the piano keyboard. There are eight white keys for every five black keys. The black keys come arranged in groups of two or three, and there are 13 keys to an octave—all Fibonacci numbers.

Not only the piano but most of Western music in general is full of these ratios. The three main scales are the pentatonic (five-note scale), diatonic (eight-note scale), and the chromatic (13-note scale). A major sixth is the mixture of C (264 vibrations per second) and A (440 vibrations per second). The ratio created is 264/440, which reduces to 3/5. Even the tempo of music can express these ratios, as in the opening of Beethoven's Fifth: bum, bum, bum, bummmmmmmmm, bum, bum, bum, bummmmmmmmmm.

The chemical makeup of pulses traveling along nerve fibers, the different possible histories of the electron, the growth curve of a city's energy needs, studies of poultry inbreeding in Australia—all use formula based on Fibonacci numbers. Astronomers have found them in formula used to predict the distances of the moons from the planets Jupiter, Saturn, and Uranus. They show up in ocean waves and galaxy distribution. Streams flow into rivers like tree branches in reverse, cutting Fibonacci numbers across the land.

One of the most interesting places they show up was found in the 1930s by Ralph Nelson Elliot. He was studying the Dow Jones Industrial averages and noticed the rising and falling of prices mimicked Fibonacci numbers. From this came the important Elliot Wave Principle, and it's used to spot trends in the market. Elliot noticed the pattern: five waves of up prices, three waves down, for a complete cycle of eight waves.

So they show up a lot. So what? Why do phylotactic ratios come in Fibonacci numbers? Do they represent some mathematical structure to nature? Are you surprised that the Fibonacci ratio gets closer and closer to .618034, the golden ratio?

Pascal's Triangle

This triangular array of numbers was known before Pascal, but he made it famous and so it bears his name. It was of interest to very early Chinese mathematicians, and may be of interest to you. Since a Pascal Triangle is easy to construct, you can start by making your own. Write the digit 1 at the top middle of a piece of paper. This meager beginning—and a strict rule on creating new rows for the triangle—produces a numerical arraignment with fascinating properties.

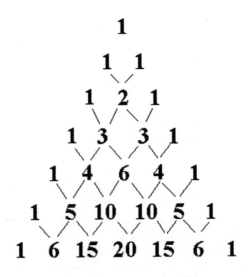

Pascal's Triangle

Each successive line is made by adding the numbers of the previous row. Since the first row contains only 1, there's no addition to make, and the next row is simply 1, 1. Since there's no number on the far right and left of the triangle, the outside numbers of every row will be 1. Since there's nothing to the right of the leftmost 1 and nothing to the left of the leftmost 1 on any row, you can think of this as 1 + 0, which generates the 1s that cascade down either side of Pascal's Triangle. Within the triangle though, the numbers will increase in size and quantity.

The third line is made by adding 0 + 1 (there's nothing to the left of 1, so add 0); 1 + 1 is 2, and this is the middle number of the third row; the last number is, again, 1 (1 + 0). Continuing in this way, you can construct a Pascal Triangle as large as you like.

One downward diagonal gives the natural numbers. The other diagonal gives the triangular numbers. Most important of all are the horizontal lines. Formulas of the form $(a + b)^2$ are expanded into statements of the form $a^2 + 2ab + b^2$. When the exponent is large, the coefficients of the second statement are not so easy to find. The horizontal rows of Pascal's Triangle give the coefficients for binomial formulas of any degree. For instance, a formula like $(a + b)^6$, can be expanded to this formula, $a^6 + 6a^5b + 15a^4b^2 + 20a^3b^3 + 15a^2b^4 + 6ab^5 + b^6$. The coefficients for a sixth-degree binomial are 1, 6, 15, 20, 15, 6, and 1, and these are the numbers on found on the seventh row of Pascal's Triangle.

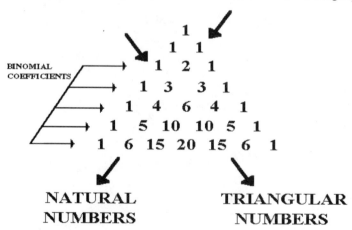

Pascal's Triangle

In fact, if you need to know the coefficients for a formula of the form, $(a + b)^{23}$, just create a 24-row Pascal's Triangle. Before you begin, make sure you have a big piece of paper—and start at the top. The usefulness of Pascal's Triangle dimmed when Isaac Newton discovered the binomial theorem, which calculates coefficients without having to build a triangular construction.

The sum of all numbers on any particular row of Pascal's Triangle reveals an interesting feature. The sums of the rows of Pascal's Triangle are 2, 4, 8, 16, ..., and these are also the solutions to the series 2 to the nth. If you need to figure the answer to 2 to the 275th, just construct a Pascal's Triangle of 276 rows. The 1 at the apex of the triangle could only be 2^0, which may not make intuitive sense; it is, after all, 2 multiplied by itself 0 times. It's value is 1, though, or at least mathematicians translate it as 1, and that is also borne out by Pascal's amazing triangle.

$$1 = 2^0$$
$$1 + 1 = 2 = 2^1$$
$$1 + 2 + 1 = 4 = 2^2$$
$$1 + 3 + 3 + 1 = 8 = 2^3$$
$$1 + 4 + 6 + 4 + 1 = 16 = 2^4$$
$$1 + 5 + 10 + 10 + 5 + 1 = 32 = 2^5$$
$$1 + 6 + 15 + 20 + 15 + 6 + 1 = 64 = 2^6$$

Pascal's Triangle

Another triangle takes an interesting pattern. Start in the same way as Pascal's triangle, but create each new row from the difference of the numbers above it. The first few rows are:

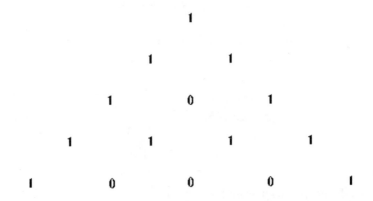

What if you continue to build the triangle on this difference formula?

The Math You Might Have Learned

Y ou knew I'd get around to these topics sooner or later. Logarithms, algebra, and calculus are too important to leave out. It would be hard to think of any science that could do without these basic mathematical tools, even though most of the rest of us do. Recently even the social sciences have taken mathematical tools and shaped them to their needs, but at one time even the physical sciences did without logarithms, algebra, and calculus.

Logarithms

To do serious astronomical work, astronomers used trigonometric values like sines and cosines, which are ratios of sides and angles in triangles. The only trouble was that these values are hard to calculate. To make matters simpler, great tables of values had to be compiled before any real work could begin. People looked for a simpler way of working with sines, and Scottish mathematician John Napier labored more than 20 years to publish his method of simplifying these calculations in 1614.

Logarithm is the Greek word for "proportional numbers," so Napier used it to describe his method of calculation. Science was dealing with bigger and bigger numbers, and logarithms made calculating these big numbers easier. Almost everyone will agree that adding and subtracting is simpler than multiplying and dividing, and Napier's method translated multiplication into addition and division into subtraction. The secret lies in exponents.

When a number is squared, the little 2 at the top is the exponent. The following table shows the first 10 powers of 2:

1	2	3	4	5	6	7	8	9	10
2^1	2^2	2^3	2^4	2^5	2^6	2^7	2^8	2^9	2^{10}
2	4	8	16	32	64	128	256	512	1024
3^1	3^2	3^3	3^4	3^5	3^6	3^7	3^8	3^9	3^{10}
3	9	27	81	243	729	2187	6561	19683	59049

Exponents indicate how many times to multiply a number together, so 3^3 = 3 x 3 x 3, and 3^5 = 3 x 3 x 3 x 3 x 3. By writing the exponent out in full, it's easy to see that:

3^3 x 3^5 = (3 x 3 x 3) x (3 x 3 x 3 x 3 x 3) = 3 x 3 x 3 x 3 x 3 x 3 x 3 x 3 = 3^8.

When the calculation requires you to multiply numbers with exponents, just add the exponents. Referring to the table of powers, to multiply 27 times 2187, just translate the numbers to their exponents and add the exponents so you end up with the simple addition, 3^3 x 3^7 = 3^{3+7} = 3^{10} = 59049. If you need to divide, subtract the exponents like so:

$$\frac{19683}{729} = \frac{3^9}{3^6} = 3^{9-6} = 3^3 = 27$$

Adding and subtracting makes quick work out of multiplying or dividing big numbers, but only after you've worked out the tables. And what if you need to multiply 16 by 27? Exponents can only be added and subtracted when both numbers have the same base. And some numbers aren't available on either of our lists of powers of 2 and 3. How can these numbers be handled?

As long as the base of the logarithm is the same, adding exponents gives the same answer as multiplying the numbers. The common logarithms are calculated on base 10, so that:

1	0.0000	$10^{0.0000}$
2	0.3010	$10^{0.3010}$
5	0.6990	$10^{0.6990}$
10	1.0000	$10^{1.0000}$
20	1.3010	$10^{1.3010}$
50	1.6990	$10^{1.6990}$
100	2.0000	$10^{2.0000}$
200	2.3010	$10^{2.3010}$
500	2.6990	$10^{2.6990}$
1000	3.0000	$10^{3.0000}$

The part of the logarithm to the left of the decimal point is the characteristic, and the part to the right of the decimal point is the mantissa. The mantissa of common logarithms remains the same for numbers that differ by a power of 10, like 5, 50, or 500. The common logs for these numbers are 0.6990, 1.6990, and 2.6990. Using this feature of common logarithms, tables were created using only the mantissa of the numbers 1 through 10. Since the mantissa of 5 is 0.6990, only the characteristic is needed for the logarithms of 5,000 and 50,000. Since 10^3 is 1,000, and 10^4 is 10,000, the common logarithms for 5,000 and 50,000 are 3.6990 and 4.6990.

The first system Napier devised was actually based on the number 10,000,000. English mathematician Henry Briggs thought the logarithms would be easier to use if based on the number 10, Napier agreed, and in 1624 Briggs published the common logarithms from 1-20,000 and 90,000-100,000. Even before Briggs published his tables, other people had seen the utility of logarithms, and they developed a way to employ them that didn't depend on numerical tables.

Sliding In

Only six years after Napier published his first tables, Edmund Gunter took a logarithm scale and put a pair of pointers on it. William Oughtred literally put two logarithm scales on top of each other so they could slide, and the first slide rule was invented. To multiply 5 times 6, you just slide the beginning of the top scale to 5 on the bottom scale and read the number under the 6 on the top scale. The slide rule was so helpful for calculating large numbers that it became the most important tool for working with numbers, and remained so until the 1970s saw the introduction of the affordable electric calculator.

Algebra

Algebra is the branch of mathematics that deals with unknown quantities, or variables. The variables, indicated by x, y, and z, represent real numbers, just unknown real numbers. Using some numbers, some variables, and the usual rules of arithmetic, mathematicians construct sentences that reflect some relation between the known numbers and the unknown variables.

Sometimes these mathematical sentences express equalities, and sometimes they represent inequalities, like so:

$X^2 - 3x = 0$

$X^2 < 16$

By performing operations on these sentences, it may be possible to isolate the unknown x, y, or z on one side of the equation. At the end of the Dark Ages, science was beginning to seek answers to lots of unknowns and algebra became an indispensable tool for them. All science today uses the algebraic variables x, y, and z, but these symbols were only introduced in 1591. That was the year Francois Viete wrote an algebra book using essentially the same symbols you can find in today's algebra books. The beginnings of algebra, however, are much older.

Ancient Egyptian and Babylonian mathematicians did algebra, but they expressed their mathematical sentences in words. The great Greeks solved for unknowns, but they also put these equations into words. Not only were these "word problems" difficult to follow, the Greeks approached the problems in terms of their revered geometry. A modern algebraic sentence like: $x (x - 6) = 20$, was expressed by Euclid like so:

"If a certain straight line be diminished by six, the rectangle contained by the whole and the diminished segment is equal to 20."

No wonder the Greeks preferred to work with lines and circles.

One of the first published approaches to algebraic problem-solving is the *Disquisitions,* a 3rd-century book by the Alexandrian Greek Diophantus. Diophantus sought whole-number solutions, and even today equations with whole-number solutions are called Diophantine Equations. Diophantus didn't bother to find solutions for all algebraic equations. He was a frugal problem-solver, for as soon as he found one solution for an equation he quit. Once was enough for Diophantus.

Al-Khowarizmi's book on algebra, *The Science of Restoring and Canceling,* was the next great algebra text, but it became really great only after Fibonacci published his translation of it. That translation introduced "al-jabr" to

Europe. By now some algebraic solutions were requiring negative numbers or irrational numbers for solutions, but most people were unwilling to accept these kinds of numbers as answers. Negative and irrational solutions only gained acceptance toward the end of the 17th century.

At the beginning of the 18th century, different algebras were developed that didn't use any kinds of numbers for their solutions. In the section on logic, we'll see a very powerful algebra created by George Boole called Boolean Algebra. Variables in Boole's algebra represented classes or sets, not numbers at all. On the other end of the scale, Arthur Cayley created an algebra that had whole arrays of numbers, or matrices, as the solutions to the equations.

Another important algebra came from the Irish mathematician William Rowan Hamilton, who interpreted his variables as vectors, the combination of direction and speed. Interpreting the variables of algebra as vectors created special problems. For instance, the commutative law for multiplying (3 x 4 = 4 x 3) isn't true for vectors. Losing one of the important laws of arithmetic was not what Hamilton had in mind when he created his system, but that's what happened. Vectors don't obey the commutative law of multiplication. However peculiar the system seemed without this property, Hamilton's system described the way vectors worked. Today there are lots of abstract algebras.

In the 1600s, Rene Descartes showed how to combine algebra and geometry to form analytic geometry. Using a horizontal x axis and a vertical y axis, equations could now be graphed. Lines on the graph represented solutions for the equation when certain values were known. Descartes's system worked just as well in reverse, so—since any point on the graph can be located with only two coordinates—geometrical problems could be solved algebraically.

Calculus

If there had been automobiles in the 1700s, Descartes might have used his analytic geometry to plot a graph of an automobile trip. All you have to do is interpret the horizontal axis of the graph as the time lapsed and the vertical axis as the distance traveled. If you drove 50 miles an hour for four hours, your graph would look like so:

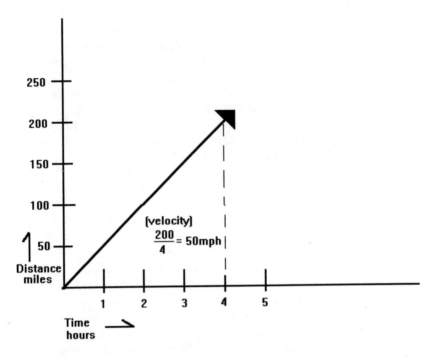

Since the velocity is a constant 50 miles an hour, the line representing velocity is straight. Geometry and algebra work perfectly for car trips, except for one little detail. Real cars on real roads don't go 50 miles per hour for four hours. Stoplights will hold you up and, on today's interstates, much of the time you may travel faster than 50 miles per hour. Even if you could keep the speedometer perfectly set at 50 miles per hour, you still have to obtain that speed from a standing start. (And hopefully you'll slow down when you arrive at your destination.)

A real car travels at a variable speed, and to account for that you need calculus. Specifically, differential calculus is needed to calculate the changes (differences) in one variable y (distance) produced by changes in another, related variable x (speed). The variables x and y are related because every value of x is paired with exactly one value of y, however long you drive. Mathematicians would say that y is a function of x, and they write this relation like so:

$$y = f(x)$$

Differential calculus examines the change in y due to a change in x, where y is a function of x. In our automobile graph, the distance is a function of speed, and differential calculus examines the change in distance that's due to a change in speed. We can also graph the car trip using the vertical axis to represent velocity. If the car accelerates from a standing start to 50 miles per hour in five seconds, the graph might look like so:

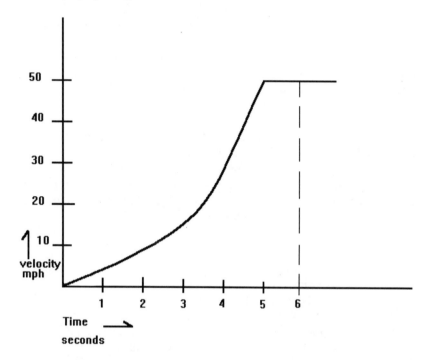

As the car accelerates, the distance is not a neat 250/5, but is the area of a figure with a curved side. From the attempts to square the circle, we know by now how difficult it is to find the area of a figure with a curved side.

Remembering Archimedes and his theory of exhaustion, we might slice this curved shape into lots of very small but easily computed shapes. The smaller the slices, the more accurate the figure we can obtain; because of this, it's sometimes called infinitesimal calculus.

This particular branch is called integral calculus. It calculates the distance, area, and volume when there are irregularities in a variable. After all, to find the speed of the car at a given point you must figure the slope of a point! So calculus figures the "limit" of lots of slopes of points, and gives a damn good approximation of the speed of the car at any instant of its acceleration. Differential calculus is useful wherever quantities are changing. Optimization problems like "How high will it go?" and "How far will it fall?" are suited to differential calculus. The limiting process is slicing these shapes into small pieces.

So calculus deals with slope-finding—or differential—calculus, and slicing and summing—or integral—calculus. At first these were completely separate studies, but Newton and Leibniz saw them to be related. They were really opposite processes, and the description of this relationship is known as the fundamental theorem of calculus. Years earlier, it was almost guessed by a few great mathematicians.

Archimedes wrote *The Method,* but a copy was only discovered recently. The unfortunate custom was to write religious texts over other writings, and for years *The Method* had stayed hidden behind mumbo-jumbo. In this work, Archimedes describes his method of exhaustion, which anticipate the concept of limits used in calculus.

Kepler described the planets' travel as an ellipse, advancing astronomy, but his method of finding the optimum shape of a wine barrel and computing its volume was very close to the methods of calculus. The credit for calculus belongs to two men, Isaac Newton and Gottfried Leibnitz, but just who deserves the solo credit is a matter of debate. Authorship credit for this powerful tool was a source of rancor for many years.

Isaac Newton

Born: 1642
Died: 1727
Hometown: Woolsthorpe, England
School Affiliation: Cambridge
Best Work: *Principia Mathematica*
Best Formula: Newton's Three Laws of Motion
Fields of Interest: Mechanics, calculus, physics
Quote: If I have seen a little farther than others it is because I have stood on the shoulders of giants.

Maybe the greatest scientist of all time, Newton built the first reflecting telescope, described the laws of gravitation, invented calculus, and explored optics and fluid dynamics. He held the prestigious Lucasian chair in Mathematics at Cambridge University, was president of the Royal Society, and was knighted by the Queen. Somewhere along the way, it is said, he napped beneath an apple tree.

Isaac's father caught the boy reading under a bush when he was supposed to be helping out—he was searching for answers early. Young Isaac became a good student at the Grantham Grammar School, and was admitted to Cambridge University in 1661. He was gifted in mathematics but didn't complete his greatest work at Cambridge. The bubonic plague struck England in 1664 and, for two years, the university was closed. Back home, Newton spent the two plague years thinking about gravitation, fluxions (calculus), and the spectrum of light.

All his life he was surprisingly reluctant to publish his findings, but at the request of astronomer Edmund Halley he wrote the *Principia Mathematica* during the period 1684-86. In this great work, he described his theory of gravitation and laws of motion. He proved all of Kepler's laws from his own, and so moved Kepler's work into theoretical form instead of just observational deductions.

About 1693, Newton learned that calculus was being embraced by mathematicians and scientists all over Europe, and it was the calculus developed by Gottfried Leibnitz! Soon accusations of plagiarism were made, and a bitter feud developed between England and Germany, both seeking to establish their man as the sole inventor of calculus.

Leibnitz and Newton both had workable systems, but the German's calculus was far easier to use and didn't depend on the concept of fluxions, or flowing quantities. National pride was strong, and besides defending their man at every turn, the British turned their back on the handier methods of Leibnitz. So strong was their support for Newton that British mathematicians did little but defend him for almost 100 years.

For all Newton's greatness as mathematician and scientist, the British government rewarded him by placing him in charge of the mint. He represented Cambridge University in the British Parliament, and was knighted by Queen Anne in 1705. He died at 85 years of age and is buried in Westminster Abbey.

Infinitesimals

The proprietary concerns about calculus were serious squabbles, but other people were expressing doubts about calculus that had nothing to do with discovery credit. These people objected to some of the descriptions mathematicians were using to talk about calculus.

After all, calculus involved processes like infinite limits, or infinitesimals. The mathematics worked, but mathematicians treated continuous change, like acceleration, as if it were an infinite series of infinitely small step changes. It wasn't precise what it was you were talking about when you talked about an infinitely small step.

Newton called the infinitesimals fluxions, and his definition of them is almost religious. A fluxion isn't nothing, but it isn't much since it's infinitely small.

Bishop George Berkeley thought the use of this notion of fluxion was a departure from the clinical scientific method, and he was tired of science being held up as more certain than religion. How could mathematicians look on religion as scientifically unsubstantiated claims when mathematicians used such vague notions in their calculations? Berkeley wrote about fluxions, "They are neither finite quantities, nor quantities infinitely small, nor yet nothing. May we not call them ghosts of departed quantities?"

Berkeley accepted the usefulness of calculus, and the valid results of the method, but he couldn't abide mathematicians who claimed their subject was more clearly defined than religion. In his *Discourse Addressed to an Infidel Mathematician,* he wrote, "He who can digest a second or third fluxion, a second or third difference, need not, me thinks, be squeamish about any point in Divinity."

Odds Are

"I must complain the cards are ill-shuffled, till I have a good hand."
— Jonathan Swift

What are the odds that you'll be struck by lightning, or give birth to twins—and what can you do about it, anyway? Do you really have a chance of winning the lottery? Is there such a thing as a lucky streak, and is it mathematical? Or, when it comes to gambling, is it like the song says, "When you're hot you're hot, and when you're not, you're not"? Face it—when you come face to face with probability, you gotta ask yourself just one question: "Do I feel lucky?"

Well do ya, punk? Lots of our ancestors did. Dice are among the items found in the ancient tombs of Egypt and Asia. These early high rollers undoubtedly confused chance with fate and blamed the gods—or blessed them, depending on whether they rolled their point. Though dice are ancient, the mathematics of chance and probability were practically nonexistent before the 1600s.

Notions of chance and probability violated the accepted order of the universe. Mathematics contributed mightily to this belief in universal order, and it probably seemed unnatural to use math to look at chance events—even if you did believe there were chance events. By the 1600s, better, fairer dice were being made, and this, too, may have been a factor in the study of probability. Many early dice were carved or painted bones, and the same ancient scrolls that describe dice and dice games make mention of loaded bones, as well.

Galileo was certainly familiar with the valid statements mathematics was beginning to make in the sciences. Always bucking the system, he wondered if mathematics could make valid predictions at the craps table. In a treatise, he set out some basic probability theorems taken from questions asked of him by a gambling acquaintance.

This Italian nobleman wondered why, when he tossed three six-sided dice, the combined number was 10 more often than 9. Galileo figured that since there were six ways each of the three die could come up, there were 216 possible combinations (6 x 6 x 6). Since out of these 216 possible combinations 27 add up to 10 and only 25 combinations add up to nine, it's likely that 10s will come up more often than 9s.

The 16th century Italian mathematician Cardano published a gamblers manual in the late 1500s, and his work and Galileo's thoughts were markers for what was to come. But the beginning of a solid mathematical theory of probability is found in the work of Blaise Pascal (of triangle fame) and the most talented amateur in all mathematics, Pierre De Fermat. Probability wasn't hammered out in some lecture hall or science lab. This potent field of mathematics had its real beginnings in a correspondence between friends.

Letters On a Dice Game

In 1653, the Chevalier de Méré had a problem, and it was a problem about dice. He'd been winning steadily by betting even money that out of four rolls of a single die he would roll a six at least once. As a good cavalier should, he tried to parlay this winning strategy. He bet even money that out of 24 rolls with two dice he would roll at least one double six. Now de Méré was losing.

Problems such as this require immediate, professional help, so the confused gambler turned to Blaise Pascal, renowned philosopher, physicist, and mathematician. He was a good person to ask. Pascal thought about it, and jotted off his ideas to his friend Fermat. The correspondence between these two laid the foundation for the field of probability.

The Chevalier's second bet is at odds with his odds, so to speak. Rolling a six once in four rolls of one die is 3.549 percent more likely to happen than not rolling a six in four tosses. This small advantage accounts for the pots he won with the one-die bet.

With the second bet, the gambler's reasoning had been off—but not by much. It's more likely than not that you don't roll a double six in 24 tries with two dice. Mathematically, you're more likely not to roll a double six at all in 24 tries by a percentage of 1.27 percent. Pascal discovered that the odds would favor the bet by .85 percent if the bet was extended to 25 tries. The Chevalier was so close.

Another rule of probability that can be derived from this is that gamblers don't read Pascal. In 1952, a New York City gambler named Fat the Butch tried to win a huge pot by betting that he could roll a double six in only 21 rolls—worse odds even than Chevalier de Méré by 20.45 percent!

As Pascal and Fermat knew, the laws of probability will only work for repeatable events with definite outcomes. The "laws" can be broken often when only a few events are tested, like tossing a coin. You know that for a coin to come up heads, the chances are 50-50, yet on three tosses you could get heads every time and it wouldn't break the "law." Still, probability says that for thousands of coin tosses—better yet, millions—the percentage of heads and tails will be the same. How does this work?

Sample Space

The term for the set of all possible outcomes of an experiment is sample space. For tossing a nickel, the sample space consists of two outcomes: {heads, tails}. Rolling a pair of six-sided dice creates a sample space of 36 ordered pairs. An event is a well-defined subset of the sample space. For instance, if the sum of the dice toss is 6, that event consists of 1 of 5 outcomes; (1,5), (2,4), (3,3), (4,2), (5,1).

A ratio expressing the chance or likelihood that a certain event will happen, given the number of possible outcomes of an action, is probability. If an event is certain, the probability is 1; when an event is impossible, the probability is 0.

The probability of A,P(A), is expressed mathematically by the formula:

$$\text{Probability of A} \atop P(A) = \frac{\text{number of outcomes of interest}}{\text{number of possible outcomes}}$$

$$\text{Probability of a coin being heads} \atop P(A) = \frac{\text{number of outcomes of interest (heads) (1)}}{\text{number of possible outcomes (2)}} = .5$$

As you can see, when the event in question is impossible, the numerator of the formula is zero—no matter how many possible outcomes are represented in the denominator of the formula. Probability can be figured for all kinds of events, and the probability of an event not happening can be figured just as well.

Probability can help even when you want the probability of two events. The probability of A or B is written mathematically like this: P(A or B) = P(A) + P(B) - P(A + B). This probability formula works only for two events that can occur at the same time, like having red hair and being a man. Mutually exclusive events, events that can't happen at the same time, are expressed by the simple formula: P(A or B)= P(A) + P(B), because both A and B must happen.

That's Odds

Odds are different from probability. The odds that an event will happen is the ratio of the probability of that event to the probability of its not happening. Therefore, while the probability of rolling a 4 with one roll of one die is one-in-six, the odds of it happening are 1/6 divided by 5/6 (the probability that your roll won't be a 4). The odds of rolling a 4 is one-to-five.

If you play the ponies, you know that the odds at a racetrack are usually expressed in terms of the odds against something happening. A five-to-one ratio on Sea Biscuit in the fifth race, for instance, means five to one against him winning.

Another probability idea is the mathematical expectation, or the expected value of a chance. The mathematical expectation of an event is the probability of that event times the value of the event. In this case, the value of an event might be money or points in a game. All fair games have a mathematical expectation of zero. Therefore, when the probability of two events is 1/2 or 50-50 for both but the value of heads is two dollars (while the value of tails is one dollar), the mathematical expectation of heads is greater.

Blaise Pascal's groundbreaking work on probability through gambling didn't diminish his strict religious beliefs. In fact, he used his theory of mathematical expectation to argue for a belief in God. Known as Pascal's Wager, the argument goes like this: Belief in God promises a heavenly reward. This reward from God must surely be an infinite reward, and so, no matter what the probability that God exists, that figure, no matter how small, is going to be multiplied by infinity (the reward). Therefore the mathematical expectation of belief in God is infinite. To Pascal, believing in God was a sure thing.

The Gambler's Fallacy

Flip a coin. What are the odds that it will land heads? Fifty-fifty, of course. And if you flip it 10 times in a row, and every toss comes up heads, what are the chances that you'll get heads the next time? Probability tells us that every time the coin is flipped there's a 50-50 chance it will come up heads. The chance that the coin will be heads on 11 straight flips is much lower than 50-50 (instead of .50, it's 0.00048828125), but for each individual toss there will always be a 50-50 chance.

This is sometimes referred to as the gambler's fallacy, for mistaking this aspect of probability is an easy thing to do—and an easy way to lose. If the roulette wheel stops on black 10 times in a row, what are the chances it will be black the next time? The hopeful gambler forgets that every spin is an individual spin, and bets heavy on the red, though the odds of red turning up are really no better than 50-50 on any spin.

The gambler thinks the bet is whether black will come up 11 times in a row, just because it's come up 10 times already. As a matter of fact, if the house offered that bet—that black wouldn't come up 11 times in a row—the payoff odds would be very low. But in the swirl of the game (and after a few watered-down house drinks), the gambler's fallacy seems right.

Paths

Probability outcomes can be illustrated using paths that branch at every possibility. The chart for flipping a coin is shown below:

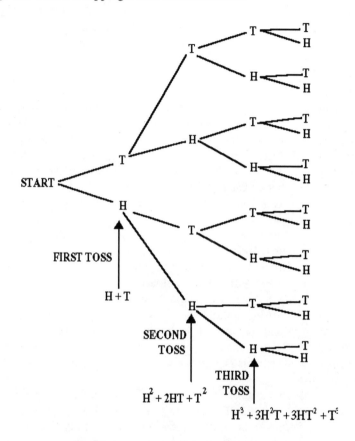

Each time you flip the coin there are two branches, heads or tails. You can trace any outcome from the pyramid path of a single coin toss, thereby tracing all possibilities of flipping a single coin. All possible combinations can be seen in the binomial theorem.

Randomness

Other actions are harder to predict than the drawing of a card, or the rolling of snake-eyes. Such processes include the Stochastic processes, which evolve over time, like Brownian motion.

With the discovery of Brownian motion in the middle of the 19th century, people tried to explain the seemingly random behavior of microscopic particles. That century's growing young science of heredity was taking from and adding to probability and statistical math in a very healthy exchange of "pure and applied" maths. But the axiomatic system of probability would wait until the 20th century.

Urn Analysis

Urn probability takes another approach to predicting information that is empirically out-of-reach, and has served us nicely in polling procedures and medicine. Suppose a barrel contains lots of colored balls. By picking several (how many?) out of the barrel, the trick is to predict the contents of the entire barrel—or, at any rate, learn *something* about a well-mixed barrel without having access to the whole barrel. One can figure intuitively that the larger sample you make will give you better odds at guessing the true contents of the urn.

Born: 1601
Died: 1665
Hometown: Beaumont-de-Lomagne, France
School Affiliation: Toulouse
Best Work: His thorough correspondences
Best Formula: $A^N + B^N = C^N$, but only when $N = 2$
Fields of Interest: Probability, number theory

Pierre De Fermat

Quote: I have discovered a truly marvelous demonstration, which this margin is too narrow to contain.

Unlike most mathematicians, Fermat worked for a living. That is, he didn't teach. He was a lawyer and son of a leather worker, but he wasn't too busy to do plenty of mathematics in his spare time. From the results he obtained, he pursued this hobby with vigor, and is undoubtedly the most accomplished amateur mathematician in history.

The government job he performed most of his life was exacting, detailed work, and he brought the same approach to his mathematics. Fermat contributed to several fields, but never published much during his lifetime. Like the probability theory he developed with Pascal, each of his contributions bore the mark of originality. Fermat is forever linked with one formula in particular.

Most of Fermat's work is contained in letters to his

friends. He sometimes scribbled in books as he read, and the quote above was written in the margins of a book by Diophantus. Fermat's "marvelous demonstration" would have proven that only square numbers solved equations like $A^N + B^N = C^N$. No numbers could be substituted for A, B, and C so that $A^Q + B^Q = C^Q$. Many people tried to reproduce Fermat's alleged proof, and Fermat's last theorem was by far the most famous unsolved problem in mathematics, at least until 1993. Though complicated and indirect, a proof has been offered that looks like it will hold up, which would make Fermat's theorem the most famous *solved* problem.

Going Out a Winner

Chance is now blamed for things that once were blamed on evil spirits, or on the devil's work.

Is probability only a measure of uncertainty? Math can't always predict a winner, but probability theory can say much about the outcomes of repeated experiments, as long as the basic conditions remain the same, like tossing dice at a reputable house. Probability needs this repeatability of experience to really get moving; things average out, but only after a long history.

In a life-and-death test of the theory of probability, in 1954 more than 2 million American children took part in a giant urn experiment. Polio was a killer on the loose, and a vaccine was at hand, but it was untested on humans. To test the drug completely would have meant a lengthy delay, and many children would certainly have died in the meantime. Employing a giant urn-type probability experiment, the vaccine was tested on the smallest number of children necessary to test the vaccine's success. The greedy pursuits of high-stakes gamblers spurred the theory of probability, and the theory passed its toughest test of all when it was used to confirm the potency of the Salk vaccine.

TRY'EM: Semi-Sure Bets

The trouble with many gamblers is that they confuse the odds with the payoff. Here are some sure bets that you can try without marking a deck of cards or loading dice.

BIRTHDAY GAMBLE. Odds are that in any room of more than 23 people two people will have the same birthday. It won't always work, but the odds are on your side. Unless you know that, however, the probability seems higher than it really is.

What you want to say is some other mumbo-jumbo about how you can sense people in the room are connected by their birthdays. Bet the room that two people present have the same birthday.

Be careful how you word the challenge. You don't want to bet that someone will have the same birthday as some individual in the room—those odds are very much against you. You only predict that two people will share a birthday. The more people over 23 you have, the better your odds get, so really lay it on when they're heavy on your side.

DOUBLE CARD DECK FLIP UP. Take two decks of cards and shuffle them. Tell everybody that you can force the decks to produce the same card somewhere, and begin turning over the cards, one from each deck, one at a time. You pick from one deck, your subject from the other deck. Tell everyone that by ESP you'll force the same card to be turned over at least once as you turn over the cards in the deck.

This is one of those tricks that should work most of the time. The odds are in your favor that somewhere along the line the same card will be turned from each deck, but if it does, keep going. If it happens twice, or three times, your "powers" will be that much more impressive. Make sure to remove the jokers, and make sure both decks are complete to keep the odds favorable for this bet.

Blaise Pascal

Born: 1623
Died: 1662
Hometown: Clermont, Auvergne, France
School Affiliation: Self-taught
Best Work: *Pensées*
Best Formula: The properties of the cycloid
Fields of Interest: Probability, geometry, atmospheric pressure
Quote: Is probability probable?

Blaise Pascal was an amateur mathematician of great renown. He was a scientist, theologian, philosopher, and religious fanatic. From a correspondence with Fermat he began the theory of probability, and it's Pascal's name we associate most closely with the famous arithmetic pyramid, Pascal's Triangle. He built a hydraulic press, discovered the cycloid, and constructed the first mechanical calculator.

Pascal suffered poor health his whole life. When he was

a boy, his father was afraid that Blaise was too bookish and frail. The boy did read a lot, so when he started asking about geometry his father forbade him to study it. On the theory that Blaise would find something else to interest him, his father took his books away. But Blaise thought hard about geometry—even without the books. He so impressed his father with his own discoveries about plane geometry that even his father agreed that the boy had a gift. His father returned the books and let him study at his own pace, and Blaise Pascal's pace was startling.

By the age of 14, he had met and impressed Father Mersenne, who held mathematical meetings in Paris. Mersenne would later be a sort of postmaster for sending letters from Pascal to Fermat and other mathematical minds of the day.

Despite the success Pascal derived from his mathematics, his fervent religious beliefs colored much of his thinking. So anti-Jesuit was Blaise Pascal that it got in the way of his study of mathematics and probability. Considering that probability grew out of the needs and curiosities of gamblers, it's surprising that Pascal practically created this field of mathematics, but he did.

A known gambler, Chevalier de Méré wanted to know how to split a pot if the game couldn't be continued—like if it was raided, maybe. Pascal's answer led to more considerations about probability, and the number of ways different things could be combined.

People like Pascal gave the Renaissance Man a good name. He also built a calculating machine, and he worked out the basic ideas of air pressure.

Shaping Up

Mathematicians developed tools for processes formerly considered "non-mathematical" like gambling, but they also looked more critically at some of the long-established mathematical tools. In the early 1800s, mathematicians began to seriously question Euclid's geometry, the very bastion of established mathematical systems. Euclid had defined mathematical thought through his geometry, and *Elements* was the hallmark of the mathematical "way." He had sewn up the subject of lines, triangles, etc., practically publishing the entire field of geometry signed, sealed, and delivered. Yet long before the 1800s people were aware of a nagging problem.

There was something funny about Euclid's fifth postulate. It described parallel lines, but people had long thought it was ill-formed. It wasn't that Euclid was wrong, only that his wording wasn't right. But how could there be something wrong with Euclid? The giant monuments of the past were erected on his system of geometry!

Pleading the Fifth

Many people had already vindicated Euclid—or said that they had. It was worth any effort to clear up this difficulty, since *Elements* had been the heart of nearly all mathematics before the 17th century. *Elements* practically described the mathematical "way." When Descartes showed how to combine geometry with algebra, this was the first essential change to Euclid's treatment of the subject, a change that occurred almost 2000 years later.

If you refer back to the five Euclidian postulates listed in Chapter Four, you can see the trouble at once. Each of the other postulates is sound, simple, and certain—and short! The fifth is wordy and full of "ifs," and its size alone makes it stand out from the others. If the fifth postulate could be proven from the other four, you could remove it from the set of axioms altogether. Proving that it followed from the others would make the postulate a theorem, and there are a lot of wordy theorems. Short of that, perhaps it would be better if the fifth postulate could be more elegantly worded.

A lot of mathematicians thought it could be better, and they went to work on a new version of the fifth postulate, but the rewording effort was always unsuccessful. The best of the lot was probably the Playfair axiom, named for John Playfair. It is simpler: "Given, in a plane, a line L and a point P not on L. Then through P there exists one and only one line parallel to L."

Better and shorter, but still not nearly as simple and straightforward as the others.

Vigorous attempts were made to prove the fifth postulate was a theorem, removing it from the set of axioms altogether. Mathematicians Saccheri, Lambert, and Legendre had made attempts, but none could prove the fifth from the other postulates. When Gauss took a look, he came to the conclusion that it simply couldn't be proven from the other postulate. Even Gauss's respected opinion didn't deter others from a sometimes fanatical search for a proof of the parallel postulate.

The search must have totally preoccupied Gauss's friend Farkas Bolyai. When Bolyai's son Janos also demonstrated an interest in the problem, the elder Bolyai wrote, "For God's sake, I beseech you, give it up," in a letter to his son. "Fear it no less than sensual passions," Bolyai continued, "because it, too, may take all your time, and deprive you of your health, peace of mind, and happiness in life." Sounds like Farkas Bolyai took it pretty seriously, doesn't it?

Non-Euclidian Geometry

Rather than defend the postulate, or flip out about it like Farkas Bolyai, some people decided to see what would happen if the fifth postulate was trashed. A geometry without the fifth postulate, however, is a geometry without parallel lines. The first geometry without the parallel postulate was published in 1829 by a young Russian mathematician named Nicolai Lobachevsky. Lobachevsky's "non-Euclidian" geometry took a very simple approach. If the parallel postulate could not be proven, Lobachevsky thought, why not replace it with some other postulate?

Nicolai chose to replace it with a postulate that allowed more than one parallel line to be drawn through a given point! In the early 1800s, it was hard to find anything in nature that could be described by a geometry that permitted lots of parallel lines through a given point. Euclidian geometry used the points off a flat plane as its subject; Lobachevsky's geometry takes only those points that form the interior of a circle. Within this domain, straight lines are defined as circular arcs that meet the circular boundary at right angles. Parallel lines in this space include all diameters, which obviously meet the circle's edge at right angles.

Lobachevsky's contribution is called hyperbolic geometry, and even as people slowly began to consider this strange geometry other solutions to the parallel problem appeared. That Lobachevsky's geometry described nothing in the "real" world seemed to be the only thing wrong with it. It had theorems and axioms, and, as a mathematical system of geometry, it was as valid as good old Euclid in every other way.

By this time, Janos Bolyai had totally disregarded his father's advice and written his contribution to new, non-Euclidian geometry. A father's pride took over when Farkas read his son's work. Bolyai had it published and showed it to his friend Gauss. Gauss approved, but said that he had come to the same ideas about the possibilities of different geometries already. He gave the work a nod, but not a wink.

In fact, the young Bolyai's work drew no real excitement from anyone in the mathematical community. People agreed that there were different, valid geometrical systems, and Bolyai's was one of them, but they were still just that—mathematical systems. With all its problems, Euclid's work was still the one that described the world we lived in. These new systems were only mathematical recreations. They had no bearing on the "real world."

And even these systems were linked intimately to Euclid's. In fact, these new systems were pretty much the same old system with a variation on the fifth postulate. Why make the whole thing revolve around parallel lines? Why even lines? What would an entirely new geometry look like?

Transformation geometry is a non-axiomatic system—it doesn't just fiddle with the parallel postulate, it does away with all Euclidian postulates. This system concerns itself with the properties of figures that don't change under certain motions. Rigid motions of shapes include translation, or moving a shape; rotation, or revolving a figure around one of its points; and reflection, or reversing the right/left aspects of the figure. A circle, for instance doesn't change at all under any of these operations.

These are rigid operations, because the basic shapes stay the same in terms of size, shape and distance. Other, non-rigid operations, like projection and distortion, do change the figure—sometimes drastically—but some properties remain. By stretching the lines of a triangle, for instance, the figure will still be convex, meaning it will still enclose an area. No matter how you distort a closed figure, it will remain closed until it's "cut."

In the middle of the 19th century, Bernhard Riemann presented another alternative to the parallel postulate. Riemann made a geometry with no parallel lines at all! Called elliptical geometry, this system can be thought of as the geometry of the points on the surface of a sphere. Like the lines of longitude on a globe, lines that seem parallel at the equator cross each other at both poles. Still, it was just another geometry that replaced the fifth postulate. It simply didn't have any reference to the "real world."

HYPERBOLIC GEOMETRY MANY PARALLEL LINES

ELLPITIC GEOMETRY NO PARALLEL LINES

 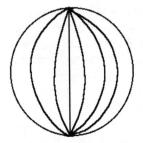

All the efforts to redeem Euclid had failed. His was not the only valid geometry. Even so, at the time of Lobachevsky and Bolyai it was the only one that seemed to describe the "real world." When the parallel postulate was finally seen as merely another assumption, no buildings fell down and all of Euclid's theorems still held. So what if there were other systems? The "real world" was still there. At least the geometry of the world was secure.

The Seven Bridges of Königsberg

In many ways, the field of topology owes its creation to a life-size puzzle found in the German city of Königsberg. In the early 18th century, of course, it wasn't a mathematical question about the geometry of the world. It was just a question of where to walk after dinner.

The residents of the German city of Königsberg enjoyed their pleasant little town, and no wonder, since it was indeed very beautiful. The Pregel river flowed right through the center of town, splitting around two small islands. There were seven bridges from the banks to the islands, and this became a popular place to walk after dinner. The puzzle for the Könegsbergians was to try to walk over all the bridges just once and not skip any bridges. The lay of the land is presented below:

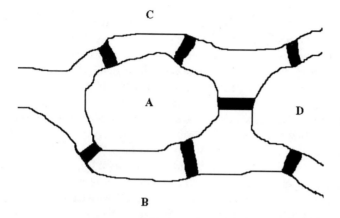

THE SEVEN BRIDGES OF KÖNIGSBERG

It remained a popular walk even though no one could walk the walk without crossing a bridge more than once or leaving one out. Most people simply thought it was impossible. Thinking something impossible is never an impediment to mathematicians, and in 1736 the amazing Euler turned his attention to the seven bridges of Königsberg.

Like any good mathematician, Euler felt no need to hike his way to the answer. Surely the problem could be solved once and for all with math. In fact, Euler's approach began by taking the pretty town of Königsberg out of the picture altogether. He drew a diagram of the bridge network, preserving the essential ingredients of the problem. His diagram looked like this:

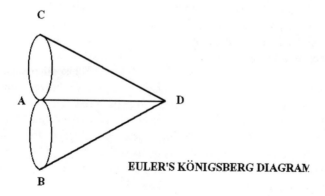

EULER'S KÖNIGSBERG DIAGRAM

The seven lines represent the seven bridges, and the points where the lines meet, called vertices, represent the land. Now the problem was easier to see. Where three bridges meet at one vertex, it was clear that some bridge was going to have to be crossed more than once. With an odd vertex, you can pass into the vertex one way and leave another, but that will still leave another arc (bridge) that must be crossed. Since this third arc must be crossed to satisfy the walk— even though you've already crossed the other two bridges—the odd vertex must be either the first or the last bridge you cross.

You could start your walk by crossing this third bridge and using the other two to pass through later, or you could pass over two bridges first and use the third bridge as your last. Unfortunately, all the vertices at Königsberg are odd. Three vertices have three arcs and the fourth (A) has five arcs. It is clearly impossible for each vertex to be the beginning or the end of the walk. The walk is impossible.

Euler created other networks of arcs and vertices and used these diagrams to discover that no network can be traveled crossing each arc only once unless the number of odd vertices is either two or zero.

Through mathematics, Euler had finally proven what the German after-dinner strollers had thought all along. The Königsberg bridge problem was solved, but the strolls continued. No mathematical proof could stop people from walking after dinner—those heavy German meals practically demanded an after-dinner walk.

Euler was so pleased with his diagram that he continued to explore interesting networks. He turned his attention to polyhedra, and discovered relationships between the vertices, edges, and faces of these solid shapes. Plato himself had found the five regular solids, and now Euler applied his "bridge" work to them. Euler did the same thing he did with the bridges and islands in the German town: He started by counting things.

Each of the solid shapes has a different number of vertices (corners), edges, and faces. All the Platonic shapes have these features in the same ratio; $V - E + F = 2$. A hexahedron, which is a simple cube, has 8 vertices, 6 faces, and 12 edges, so Euler's formula holds this way: $8 - 12 + 6 = 2$. The other shapes will also satisfy the formula; in fact, any polyhedron must have this relationship.

Unknown to Euler, Rene Descartes had noticed this relationship earlier, and so today it's usually referred to as the Euler-Descartes formula for polyhedra. The amazing thing was that the formula held for any polyhedra. You could take a polyhedra and distort it by pulling on its vertices any way you liked, but the formula remained unchanged. This relationship of vertices, edges, and faces is invariant in polyhedra.

Even when you punched a hole through one of these shapes the formula didn't change much. The mathematical relationship between vertices, edges, and faces was the same for all polyhedra, though a different number solved the equation when you considered a figure with a hole punched in it. When you take a polyhedron with a hole in it, the formula becomes $V - E + F = 0$. Even a doughnut—what mathematicians call a torus—will satisfy this equation.

Topology explores the relationships that stay the same in a shape, no matter how distorted you make it. This branch of mathematics deals with geometrical concepts like "connected," "edge," and "neighboring." Because this is the subject of topology, it's sometimes referred to as rubber-sheet geometry.

Infinity

"Infinity is a floorless room without walls or ceilings."
—Anon.

The notion of infinity confronted mathematicians thousands of years ago. Euclid's proof of the infinitude of primes describes a list of prime numbers that would "go on forever." No matter how hard each prime number might be to find, there will always be another prime. Even counting the natural numbers is an infinite process you could never finish; there's always a clear, definite way to add another number—and never a reason why you can't.

Infinity is Forever

Infinity as an unending process, something that could be continued forever, was used around 450 B.C. by Zeno of Elea when he argued for his famous paradoxes. Zeno taught that one paradoxical situation arose when you wanted to walk to the door of your house. To get to the door, you had to go halfway there first, which is pretty commonsensical. Unfortunately, Zeno went more than halfway.

He said that before you get to the halfway point (which you must reach before you get to the door), you have to go halfway there. And before you got there, you had to go halfway there, and so on. Following Zeno literally, you'll never get to the door because you'll always have to go halfway there first! It's hard to see how you could even get out of your chair, much less walk to the door.

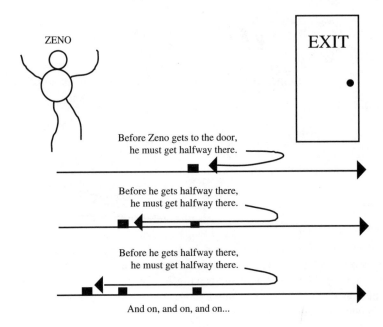

Before Zeno gets to the door,
he must get halfway there.

Before he gets halfway there,
he must get halfway there.

Before he gets halfway there,
he must get halfway there.

And on, and on, and on...

Zeno himself was wise enough to know that all one had to do to refute him was to get up and walk to the door, but it would take calculus to describe why mathematically. His paradox rested on an infinite process.

Infinity as a process is infinity as a verb, and until the 1800s it was pretty much the only way anybody ever thought about infinity. What would the noun infinity signify, anyway? How can you talk about the quantity of an infinite process? How can you talk about infinity as if it was finished? Just as the whole numbers could be either cardinal or ordinal, the cardinal number for infinity would signify the number of numbers in the infinite sequence 1, 2, 3, ... So what sense can be made of cardinal infinity?

Georg Cantor spent a good part of his life thinking about infinity and made the first real effort to treat infinity as a number, a noun. Cantor's work was original and separate from the mainstream mathematics of the day. In the mid-1800s, there were plenty of mathematical tools around, but Cantor found no slide rule or log table of any use in exploring infinity. Analytic geometry, calculus, analysis—none of the powerful mathematical fields was right for the job.

In the end, no magnificent new mathematical tool was required to muscle infinity into line, and though Cantor's theory was novel—even shocking to some—it was elegantly simple. Cantor tamed infinity with tools that had been lying around for years because, in a way, he used pebbles.

One-To-One Correspondence

When confronted with the concept of infinity, Cantor might have felt like the shepherd who couldn't count. He just didn't have the vocabulary to talk about a number as large as infinity. The shepherd had a pouch of pebbles to use because the language of numbers was beyond his understanding (probably only because he was never taught to count). Still, the shepherd could work with numbers by using his pouch of pebbles, lining up pebble for sheep, pebble for sheep, and accurately showing that there are as many sheep as pebbles. The shepherd made a one-to-one correspondence between the sheep and the pebbles.

For his "bag of pebbles," Cantor used the natural numbers, 1, 2, 3, ..., which continue forever. Anything that can be "lined" up with the natural numbers must also continue forever, and he didn't have to look far before he found an infinite number he could match off with the natural numbers. The even natural numbers, 2, 4, 6, ..., can be matched up with all of the natural numbers. Cantor saw that there was a one-to-one correspondence between the natural numbers and the even numbers. The total number of even numbers was the same as the total number of even and odd numbers put together!

In a way, you might think that there would only be half as many even numbers as odd and even put together, but Cantor showed that they were the same number. Of course, the odd numbers could also be "lined up" with the natural numbers, so there were as many odd numbers as there were odd and even put together. Once he got started, Cantor found lots of one-to-one correspondences.

$$\aleph_0 = \text{Aleph Naught}$$

The total number of natural numbers is Aleph Naught, $\aleph_0$

The total number of even numbers is Aleph Naught, $\aleph_0$

The total number of odd numbers is Aleph Naught, $\aleph_0$

There are as many even numbers as there are odd and even numbers.

There are as many odd numbers as there are odd and even numbers.

Whole Numbers	1	2	3	4	5	6	7	8	...
	↕	↕	↕	↕	↕	↕	↕	↕	
Even Numbers	2	4	6	8	10	12	14	16	...
	↕	↕	↕	↕	↕	↕	↕	↕	
Odd Numbers	3	5	7	9	11	13	15	17	...

How many primes are there? How many fractions?

How many points on a line are there?

Actually, Galileo had lined up the even numbers with the natural numbers years earlier, he just never pursued it. Cantor did pursue his search for one-to-one correspondences and found that the fractions could be lined up, as could integers (positive and negative numbers) and multiples of three, for that matter (3, 6, 9, 12, 15, ...). All that was required was to order these numbers so none were left out. The integers could be lined up this way: 1, -1, 2, -2, 3, -3, ..., which will name every integer. They can be lined up with the natural numbers. The fractions can be lined up much the same way, starting with all fractions with a numerator and denominator that add up to two, then all that add up to three, then four, and so on, like so: 1/1, 2/1, 1/2, 3/1, 2/2, 1/3, ...

All these infinite strings of numbers must have the same number of elements in the string, the same cardinal numbers, even though you can never finish the sequence. Cantor called the cardinal number of infinity *aleph naught*. Within the infinity of natural numbers, there were lots of groups, or sets of numbers, that were themselves infinite. The German mathematician Richard Dedekind even used Cantor's ideas to define what an infinite set was. He said that a set was infinite if it could be put into a one-to-one correspondence with a proper subset of itself. So there were lots of infinite sets within the infinite set of natural numbers. Doesn't this just mean that infinity is infinity?

Even Bigger

Georg Cantor didn't stop his investigation of the infinite after his discovery of aleph naught. Once he started comparing sets, he couldn't stop. He found it easy to find finite sets of numbers that couldn't be lined up with aleph naught, but Cantor was looking for an *infinite* set that couldn't be lined up with aleph naught.

Cantor tried to make a one-to-one correspondence between the natural numbers and the real numbers. Real numbers include all the rational and irrational numbers, and each real number can be represented by an infinite decimal. Suppose someone had made a one-to-one correspondence between the real and the natural numbers. The arrangement might look like so:

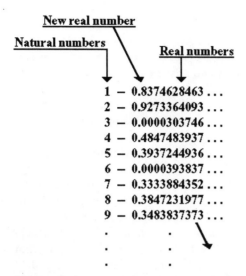

By changing each number on the new number line, we will create a real number not on the list. The new number will differ from the first number in the first digit, from the second in the second digit, from the third in the third digit, etc.

It would list all the real numbers by their infinite but distinct decimals and line them up against the natural numbers. To his surprise, Georg Cantor found a "pebble" he couldn't line up. By taking a diagonal slice through the supposed list of real numbers and changing the first digit in the first number, the second digit in the second number, the third digit in the third number, and so on, you can build a new number that's not on the list. In fact, it will differ from the first number in the first digit, the second in the second digit, and so on. There was at least one rational number more than aleph naught, because there was a way to build one. For any list of real numbers, Cantor showed how to find one that wasn't on the list—an "extra pebble." There was always a number left off the list, so the real numbers must have a total that, somehow, is bigger, or "stronger," than aleph naught. The real numbers must have a cardinal number bigger than aleph naught, and Cantor called it aleph one.

Not everyone thought Cantor should be throwing infinity around so freely. What does aleph one really mean, since you can't even finish the sequence you started with, the infinite natural numbers? Later, some of the greatest mathematicians of the day would hail Cantor's original work, but at first it drew lots of criticism. Leopold Kronecker objected most often, and some say most viciously.

An Infinite Argument

Obviously Cantor didn't construct any of his infinite sets. Not even the smallest, aleph naught, could be finished, but Cantor's theory of the infinite showed precisely how to construct infinite sets, how to work with them, and how to draw conclusions about them. To his critics, Cantor's work was far too theoretical, since in practice it was impossible to finish creating even one of Cantor's infinite sets. It was this lack of construction that formed the core of Kroenecker's attack on Cantor and aleph one.

Kronecker thought it was unproductive to treat infinity as an accomplished task, a total. How can you talk about the sum of an infinite series in any meaningful way? To be meaningful for Kronecker, a number had to be created in a precise, finite number of steps. Even more distressing to these "constructionists" was not what Cantor said about aleph naught, but what he did with it. He multiplied it by itself, added and subtracted it; he treated it like other numbers. Aleph naught times aleph naught equals aleph naught. Cantor did the math without ever completing the construction of the numbers he multiplied!

These transfinite cardinal numbers, the alephs, pushed the notion of infinity beyond just a never-ending process. It was somehow disturbing to think of transfinite numbers, levels of infinity, numbers "bigger," "stronger," "more inclusive," than the number of natural numbers. Another surprising result of Cantor's work is the discovery that there are the same number of points on lines no matter how long they are. A line of one inch has as many points as a two inch line. It is another example of lining things up to demonstrate a one-to-one correspondence, as shown in the figure below.

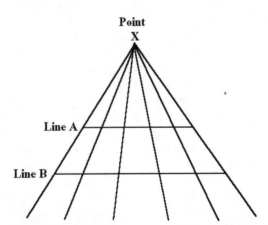

As the line from Point X sweeps through every point on
Line A, it also sweeps through every point on Line B.
There are as many points on the short line as there are
on the long line.

As Hilbert described it, it was a "paradise of infinity." There was ordinary aleph naught, the number of the natural numbers. Next aleph one, the real numbers, or the number of points on a line. Aleph two is a number bigger, or stronger, or more inclusive than even aleph one. Aleph two is the number of all geometric curves! Nobody has described a set so big it can't be lined up with aleph two. Three infinities is a lot to think about, and, as Cantor said of his own work, "I see it, but I don't believe it."

Georg
Cantor

Born: 1845
Died: 1918
Hometown: St. Petersburg, Russia
School Affiliation: University of Halle
Best Work: *Mengenlehre* (theory of sets)
Best Formula: aleph naught x aleph naught = aleph naught
Fields of Interest: Arithmetic, infinity, sets
Quote: The essence of mathematics resides in its freedom

Georg Cantor was born in Russian, lived most of his life in Germany, and died in an insane asylum. His work on transfinite numbers and set theory in the late 1800s is considered some of the most original mathematical thought ever. As a boy he showed a talent for mathematics, but his father urged him to go into engineering. Cantor would have heeded his father's wishes, but without math his mood swung so low that his dad finally allowed his son to become a mathematician.

Once he obtained his father's permission, Georg Cantor went to school at the University of Berlin. He spent a year at Gottengen, and was awarded his Ph.D. in 1867. At Berlin his teachers included Weierstrass, who had placed the calculus on firmer footing than either Newton or Leibnitz. Another teacher was Leopold Kronecker, and it was Kronecker whose attacks on his theory would haunt Georg Cantor and, according to Cantor, prevented him from receiving a professorship at Berlin. Cantor did get a professorship, but at the University of Halle, which was not nearly as prestigious as the school at Berlin.

Cantor's theory of infinite sets attracted few supporters when it was first published, and quite a few mathematicians recoiled at the surprising results of transfinite numbers. No one was a more vocal opponent than Cantor's former teacher, Kronecker. Kronecker objected to the way Cantor treated the infinite as a finished total, and Cantor reacted to this criticism very badly. He doubted his own work, and seemed totally unable or unwilling to defend himself. He was known for a bad temper, and no doubt had problems other than Kronecker's criticism, but the attack on his work was more than he could stand. He checked into an asylum for the mentally ill.

For the rest of his life, Georg Cantor would be in and out of mental hospitals. While he was out, he would deliver more novel mathematical work, but it would be the early 1900s before his work would become generally accepted. The logician Bertrand Russell would praise Cantor's work in the highest terms, and David Hilbert defended him, saying, "No one shall expel us from the paradise which Cantor has created for us." Nevertheless, Georg Cantor died in a mental hospital in 1918.

Hilbert's Hotel

Talk about leaving the light on for you. You won't find a more hospitable home away from home than Hilbert's Hotel. No one has ever been turned away. In the late 1800s, the great German mathematician David Hilbert built this hotel from materials he had on hand, and what Hilbert had on hand was Cantor's numbers.

Hilbert built a wonderful hotel that makes reservations obsolete, and there are no No Vacancy signs. At the Hilbert, you see, they have an infinite number of rooms—all with cable. When a *very* large convention came to town that consisted of an infinite number of participants, (it was a convention on The Ramifications of Infinity), they all stayed at the Hilbert. For the first time in memory, the Hotel with infinite rooms was full, and that's a lot of pillow mints!

Later that night, another conventioneer arrived in town and, wanting to join all his friends at the Hilbert, walked to the lobby desk and asked for a room. Since each of the rooms at the hotel was occupied, the desk clerk became flustered. Turning a guest away might ruin the hotel's reputation, but the clerk remembered what David Hilbert had told him to do if this situation ever came up. The guest signed the register, and the bell man showed him to his room. How did he do that?

No Vacancy takes on a different meaning when the hotel has an infinite number of rooms. The late arrival is given room No. 1, and the occupant of room No. 1 is moved to room No. 2. The occupant of room No. 2 is moved to room No. 3, the occupant of room No. 3 to No. 4, and on, and on, and on. It may seem peculiar at first, but this procedure is firmly rooted in Cantor's mathematics of infinity.

As a matter of fact, Hilbert's Hotel would be able to accommodate even another infinite number of conventioneers if they showed up. Practically speaking, though (if it's not too late), it's obvious that somebody will always be in the hallway, moving from one room to the next. Even so, no one is ever turned away from Hilbert's Hotel, and instead of a No Vacancy sign it flies several Buses Welcome banners.

Logic

"God is love, love is blind, therefore Ray Charles is God."

— Anon.

Logic is a lot of things. It is right think-
ing, it is sharp reasoning, and it is good old common sense. Sort of. Take the
very speculative argument above. Even if Ray Charles was God, that argument
wouldn't prove it, would it?

Few texts in the history of mathematics have as logical an approach as
Euclid's *Elements*, but *Elements* contained the funny parallel postulate, the
postulate that crashed the hopes of those who saw it as a pinnacle of truth. If
logic is right thinking, surely we should be able to develop a formal system of
right thinking. If logic really works, we should be able to plug information into
a logical system, perform a few calculations, and separate the good arguments
from the bad. Why can't we?

A good auto mechanic uses more logic daily than most college profes-
sors. The car won't start. Maybe it's the battery; the battery checks out. Maybe
it's the cables, cables check out. Maybe its the starter, etc. It just wouldn't serve
the purpose to check the starter connections until you knew whether there was
any juice in the battery.

Shade-tree Logicians

Mechanics think like this all day, and the good ones don't appeal to logic books
to verify their procedures. Right thinking is right thinking, and though the mechanic may not be able to put his thought processes into mathematical symbols, that could be done. And once translated into symbolic logic, the mechanic's
valid arguments would stay valid for any interpretation, not just batteries and
starters.

Logicians study the forms of arguments in a pure, non-interpreted, symbolic way, specifically avoiding discussion of an argument's content or meaning. Logic looks at the form of the argument and analyzes it, quite apart from
any of the meanings that can be attributed to symbolic statements. After all, if
the argument itself is invalid, what difference does it make what you're talking
about?

Archimedes may be rightly considered the first Greek mechanic, since so many of his inventions were mechanical. But in the sense of mechanics being good logicians, Aristotle takes the prize. Aristotle's system of logic is made clear in his books *Prior Analytics* and *Organon.*

Aristotle catalogued the rules of thought according to their various forms, which he called syllogisms. A syllogism takes the pattern of the argument above, only without the ridiculous content. Aristotle outlined many different forms, such as:

1. All men are mortal
2. Socrates is a man
3. Therefore Socrates is mortal

Aristotle's logic is a logic of classes; the class of men is mortal, and Socrates is a member of the class of men, so Socrates must be mortal. From two premises (1 and 2), a conclusion is drawn (3), and the validity of the argument doesn't change no matter how you change the content of the argument:

1. All mammals produce live births
2. Bobcats are mammals
3. Therefore bobcats produce live births

Or:

1. All men are athletes
2. Willie Nelson is a man
3. Therefore Willie is an athlete

The conclusions are logically valid, but not necessarily true. The premises—the two statements before the conclusions—may not be agreed on by everyone, maybe by no one. But if the premises are true, then so is the conclusion. The bobcat argument is true and valid because all the premises are true and the argument is sound. In the second example, the first premise is false because not all men are athletes. It is still logically valid because, if the premise were true, then Willie Nelson would be an athlete.

Even in Willie Nelson's case the argument holds. Logic can't decide which premises are true and which are false, but if the premises are true the laws of logic will not lead to a false conclusion. There may not even be a way to test the truth of the premises, as in the following argument:

1. All angels are holy
2. Gabriel is an angel
3. Therefore Gabriel is holy

Plenty of people would argue that there are no angels at all, holy or not. The strength of logical validity is in the fact that it doesn't matter what you're talking about, because logic is focusing on how you're talking about it. The logical argument is most clearly seen when there's no chance whatsoever to get hung up on the content, as in the following argument:

1. All A are B
2. X is A
3. Therefore X is B

Aristotle collected so many different syllogisms that students used mnemonic names to keep them all straight. Aristotle's program started from basic laws taken by assumption to be true, and he proved other statements by the right application of rules of manipulation. Aristotle's three basic laws were:

1. The law of identity: A thing is itself
2. The law of excluded middle: A statement is either true or false
3. The law of non-contradiction: No statement is both true and false

These three laws and the syllogisms constituted the whole study of logic until the 17th century. Stoic philosophers and medieval thinkers worked on logical arguments and made minor contributions to the field, but Aristotle's logic dominated in much the same way that Euclid's *Elements* dominated geometry.

Leibnitz made his name with infinitesimal calculus and in the philosophy of ethics, but he also made an important attempt to contribute to the study of logic. Leibnitz thought that logic could become a universal tool for right thinking, a tool that could be interpreted according to the problem at hand. The two disputing parties would state their cases, the facts would be translated into Leibnitz's logical language, the Characteristica Universalis, and the correct answer would be forthcoming.

With a tool like this, right would always out—war might even end. Obviously Leibnitz's quest failed; his attempt is really only an outline. This prolific thinker never completed the international language of logic, and today only fragments of the Characterista Universalis exist. True symbolic logic would languish for another 100 years until British mathematicians revived it.

England Wakes

Since the smash-up with Germany over who was the rightful creator of calculus, British mathematicians had spent most of their time defending Newton. They spent little effort creating new works. The proprietary squabble over calculus became a source of national honor, and keeping the British end up left no room for much original work.

In the early 1800s, Augustus De Morgan began to make attempts to resuscitate logic and make it a part of the rigorous system of algebra. The form of Aristotle's arguments seemed quaint in light of the power of analysis, algebra, and other current mathematical systems. De Morgan is now remembered by logic students for De Morgan's Law, but his major contribution was recognizing an original attempt to systematize logic in the work of one of his friends.

In 1847, George Boole published the *Mathematical Analysis of Logic*. A more comprehensive edition, called *An Investigation of the Laws of Thought*, was published several years later. Now called Boolean Algebra, the system George Boole created looked a lot like algebra and had its own symbols and methods of solving logic problems. Just as the content of arguments could be symbolized, so could the connectives, or "operators" of sentences. By using symbols for words like "and," "or," "not," "if ... then ...," and small letters "p, q, r, ..." to represent statements that were either true or false, Boole symbolized logical arguments. Now just the bare bones of an argument was all that was left.

Here are some sample sentences of symbolic logic:

p and q	$p \cdot q$
p or q	$p \vee q$
if p then q	$p \rightarrow q$
p if and only if q	$p \equiv q$
not p	$\sim p$

Truth tables:

When the values of p, q are   The value of these logical sentences is

p	q		$p \cdot q$	$p \vee q$	$p \rightarrow q$	$p \equiv q$
T	T		T	T	T	T
T	F		F	T	F	F
F	T		F	T	T	F
F	F		F	F	T	T

Since every p and q stood for a proposition that was either true or false, the truth or falsity of logical sentences could be determined by a truth table. For the sentence p • q to be true, both p and q must be true. The sentence p v q, p or q, is true when either or both p or q are true. Sometimes the sentence p or q is not true when both p and q are true, as when you are offered pie or cake but not both. This exclusive "or" can be represented symbolically by the sentence, ((p v q) • ~(p • q)).

By manipulating the symbols according to set rules, many theorems are proven in much the same way they are in algebra. Boole had mathematicized the laws of logical thought, and made logic a part of mathematics. Soon others saw the power of a simple axiomatic system from which logical truths could be derived. If the mathematical truths could be derived from a system like this, then mathematics itself would finally have a firm foundation on which to rest. While Leibnitz and others had dreamed of making logic a part of mathematics, Boole did it. Its hard to overstate the effect George Boole made on the mathematics of logic, and even harder to overstate how minimal his academic training was. If there ever was one, George Boole was a self-taught man.

George Boole

Born: 1815
Died: 1864
Hometown: Lincoln, England
School Affiliation: local lower-class boys school
Best Work: *An Investigation of the Laws of Thought*
Best Formula: If A then B; If B then C; If A then C
Fields of Interest: Algebra, logic, sets
Quote: ...to investigate the fundamental laws of those operations of the mind by which reasoning is performed

The British occupation of India exposed the strict Hindu caste system in detail, but far less attention was paid to Britain's own caste system. Though certainly more informal—almost whispered—there were strict ranks of social status in Britain, and when George Boole was born to London shopkeepers in November of 1815 he was born into the lowest of the lower classes.

The distinguishing marks of the British upper class included a knowledge of foreign languages, philosophies, and even mathematics, but schools set up for children like George Boole didn't teach these subjects. The British didn't expect much from the sons and daughters of simple shopkeepers. George took the situation to heart, and resolved to teach himself Latin and Greek, which he did. Luckily, his father had learned some math on his own, and he taught young George what he knew. Unfortunately, the Boole family's situation declined even further when George was only 16, and to help make ends meet he became a teacher's assistant in the same lower-class school he found so inadequate as a pupil.

Boole tried to teach better than he had been taught, and in the evening he continued to teach himself more. No matter how much he taught himself, however, George knew he would never lift himself out of the lower class without a respectable position. To snobbish Londoners, his current level of teaching was no better than that of a shopkeeper, and since he couldn't afford to enter a profession he took refuge in the church.

After a short time, Boole saw that the meager life of a clergyman-in-training was not putting food on his parents' table. He decided to set up a proper school, not like the normal lower-class schools. Boole would really teach, and his school would teach everything, not just enough to keep the lower classes useful. Since he'd be teaching mathematics proper, not just enough for shopkeepers and coal miners, he decided to learn some himself.

George Boole found that he liked math, and he even got a paper published in a mathematical journal. Boole began to correspond informally with some of the leading mathematicians of the day, including De Morgan, who was excited by Boole's logic. After Boole published his logic, he obtained a professorship at a new college in Ireland, and taught there until he died.

Years of hard work and self-education paid off for George Boole. His work was recognized, even hailed, and whatever anyone thought of his social station, he had attained the highest respect from his mathematical peers. With *Laws of Thought,* this self-taught son of a shopkeeper provided the basis for all future logical study, and had, in Bertrand Russell's words, "discovered pure mathematical thought."

Boole's logical symbols looked so much like algebra that his system is called Boolean algebra. Others offered different methods of representing the laws of thought, and these methods didn't bring algebraic symbols into play. In 1881, another British mathematician, John Venn, developed a system of logic that for all the world looked like geometry. Instead of symbols and connectives, Venn drew circles to express logical relationships. Some simple relationships are described by the following Venn diagrams:

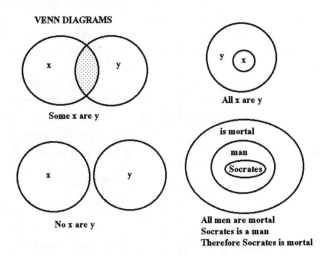

VENN DIAGRAMS

Some x are y

All x are y

No x are y

All men are mortal
Socrates is a man
Therefore Socrates is mortal

Venn diagrams are useful to demonstrate lots of relationships in a more obvious way than strings of logical symbols and connectives. You may have seen diagrams like this before in this book, since the diagram of the real numbers in Chapter Ten is a Venn diagram.

In 1896, Charles Dodgson, better known as the author of *Alice in Wonderland,* Lewis Carroll, published a book on logic. Within the book he included a game that consisted of markers and a board something like the one presented below:

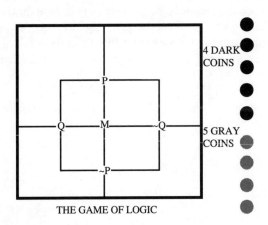

THE GAME OF LOGIC

By placing a dark marker in a cell, or on a line, you indicate that the cell or line is occupied. Light markers signify that that position is unoccupied. The M in the middle can be interpreted as some property. Take the sentence, "all men are mortal," and interpret P as men and M as mortal. The sentence can now be represented on the diagram by placing a dark counter over the P. Whether Q, or ˜ Q, the P square is occupied; that is, "All P are M." Some other examples are given below:

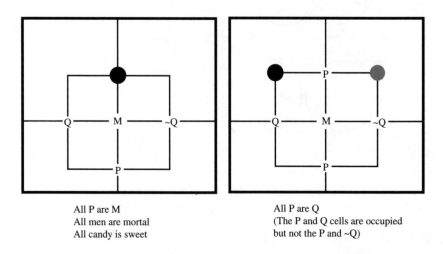

All P are M
All men are mortal
All candy is sweet

All P are Q
(The P and Q cells are occupied
but not the P and ~Q)

Beam Me Up, My Dear Watson

The spirit of logic lives nowhere more colorfully than in the adventures of Sherlock Holmes, unless it's at the science station of the U.S.S. *Enterprise* on "Star Trek." Spock and Holmes take great pleasure in the logical approach, and both make efforts to keep to deduction and close off emotion. In many cases, this purely logical approach is appealing, but it can seem foreign and robotic when viewed in the light of the fairly illogical world in which we live.

Incompleteness

N ew centuries can bring out the best in people. They're a time to reflect on accomplishments of the last 100 years and to look forward to new challenges, new hopes, and new beginnings. The start of a new century is a cause for celebration, like a higher-order New Year's.

New centuries can also bring out the worst in people. For these doom sayers, it's a time to carry "The End is Near" placards.

The people in the world of mathematics greeted the new century in 1900 by attending a convention in Paris. All the leading figures of the day were there to hear the German mathematician David Hilbert address them. Hilbert was no doom sayer, for while he did think the end was near, he was referring to the end of the remaining problems in mathematics.

There were problems remaining, that was for sure, but for Hilbert they were imminently solvable. The mathematician's toolbox was bulging with the recent results in analysis and logic, and though some of the challenges facing math were knotty, at least they were well-formed. Solutions to these problems would be the polish on a practically finished marble statue. As far as David Hilbert could see, there were 23 important questions facing mathematicians, and he challenged his peers to solve them: "Wir müssen wissen, wir verden wissen." We must know, we shall know.

Born: 1862
Died: 1943
Hometown: Königsberg
School Affiliation: Göttingen
Best Work: *Grundlagen der Geometrie*
Best Formula: Mathematical proof is also an object of mathematical research
Fields of Interest: Sets, infinity, mathematical systems

David Hilbert

Quote: No one shall expel us from the paradise which Cantor has created for us.

Hilbert's 23 problems weren't small potatoes, but each was clearly stated and understood. Once the answers to these questions were found, as they surely would be, David Hilbert figured the world of mathematics could at last be said to rest on the firmest foundation and take its place as the complete, consistent Queen of the Sciences.

Invariants and geometry, Cantor's transfinite paradise—no field of mathematics escaped Hilbert's brilliant eye. He spearheaded the formalist school of mathematics and defended Cantor's transfinite numbers against the constructionists. Hilbert was the focus of German mathematics, accompanied by his friend Minkowski, who developed a mathematics for relativity.

Hilbert headed the university at Göttingen when the very best mathematicians were there, and many were there because of Hilbert. Even Hilbert's position could not overcome the prejudice of the day, and for years he fought an outrageous battle to bestow a professorship on Emmy Noether. Noether is now known for her original work in abstract algebra, and Hilbert recognized her talent early, but at the time professorships weren't given to women. Hilbert's response was to point out that Göttingen was a university, not a bathing club.

With the 20th century, mathematics itself has grown far too large for anyone to be expert in every field, much less to dominate as Hilbert did. In many ways, he was the last total mathematician; he touched practically every field with his work.

Coming to the defense of Cantor and his infinities, Hilbert said, "No one will drive us out of this paradise that Cantor has created for us!" David Hilbert saw as well as anyone the beauty in Cantor's work, and rushed to defend it against the constructionists who would only accept proofs of things that could be constructed. New ideas often receive rough treatment, but Hilbert stayed ready to accept new fields of math, no matter their popularity.

Foundations

No one in the world was better able to address the 1900 congress of mathematicians than David Hilbert. Hilbert was at the top of the mathematical world. His contributions ranged over many fields, and he had just published *Foundations of Geometry,* a work that wrested order out of the complications brought on by non-Euclidian geometry. Since there were other valid geometries besides Euclidian, people asked, "How do we know the Euclidian system is valid?" In Hilbert's geometry, these difficulties were avoided by not defining certain concepts as physical properties.

Hilbert was so careful to avoid physical definitions of words like *point, line,* and *space* that he ended up not defining these concepts at all! The power of Hilbert's work was that it showed how unnecessary it was to define even these basic concepts. The system of geometry remains valid, never mind how the symbols might be interpreted. Hilbert proved the axioms of Euclidian geometry to be self-consistent, placing this powerful mathematical tool on firmer, more logical bedrock.

Strict foundations had already been established for most of the rest of mathematics. In 1884, Gottlob Frege published his work on the foundations of arithmetic. From a few simple assumptions, he had managed to derive a system that included all the properties of whole numbers. At this time, some people saw mathematics and logic as distinct fields, but not Frege. He began to work on a formal system of logic from which he could derive all of mathematics. Volume one of *Basic Laws of Arithmetic* was published in 1893. In Italy, Guiseppe Peano was after the same goal. His system was to embrace all of logic and most of mathematics as well. To do that, Peano needed some very powerful axioms, and they're still known as Peano's postulates.

At the mathematical congress, Hilbert challenged his peers to produce an axiomatization of physics; he asked them to show that it was impossible to solve the general seventh-degree equation by functions of two variables; he asked them which geometries had axioms most like Euclid's. And he asked them to prove that the axioms of arithmetic were consistent, something Frege and Peano were already attempting. We must know, we shall know.

Frege Gets a Shave

On the verge of publishing the final volume of his attempt to place mathematics on logical footing, Frege received a letter from Bertrand Russell that would change everything. There was a problem with logic itself. When Frege read Russell's letter, he realized that the massive work he was only days away from publishing was already obsolete. After all, what good was it to put mathematics on a logical foundation if the foundation of logic itself was shaky?

What Russell noticed was a paradox similar to the ancient Greek paradox of the liar. When someone says "I am lying," should you believe him? If the person is a liar, then he is telling the truth when he says he's lying. If he's telling the truth when he says "I am lying," he's lying, and so he's telling the truth. Somehow it seems that the person who says "I am lying" is both telling the truth and lying at the same time.

Russell's paradox is a little more entertaining. Suppose there's one barber in the town, and this barber shaves everyone who doesn't shave himself. Who shaves the barber? How can the barber shave himself if he shaves only those who don't shave themselves? If the barber is shaved by the barber, it's only because he doesn't shave himself. In either case, the barber shaves and doesn't shave himself. Of course, the whole thing is resolved if the barber has no need of shaving—perhaps she's a woman!

Russell saw that paradoxes like this could easily spring up in the logical systems offered as a foundation for mathematics. Grelling offered another paradox that deals with special adjectives. An adjective is described as autological if it possesses the property it denotes. The adjective *English* is autological since it's English, just as *Français* possesses the property it denotes. Not all adjectives are autological, like *French* and *Anglais,* which are heterological since they don't possess the property they denote.

Grelling's challenge was to divide all the adjectives into either the autological pile or the heterological pile. Adjectives like *red* and *green* would be put in the autological pile only if they were written with the appropriate color ink, making them possess the property they denoted. If you used very small lettering to write the adjective *tiny,* you could put it in the autological pile, since it would then possess the property it denotes. But the word *heterological* is itself an adjective. Which pile should it go in?

If the word possesses the property it denotes it is autological, but *heterological* denotes the property of not possessing the property it denotes. If the word doesn't possess the property it denotes then it's heterological, but *heterological* denotes the property of not possessing the property it denotes. So if *heterological* is autological it's heterological, and if *heterological* is heterological it's autological. No wonder they couldn't base mathematics on systems that could act like this.

Russell sought a way around these paradoxes, and with Alfred North Whitehead he began a massive attempt in the *Principia Mathematica.* Their system was unwieldy, to say the least; to many it was so complicated it was unreadable. Still it was hoped that by removing the paradoxes mathematics could finally have its foundation.

Born: 1872
Died: 1970
Hometown: Trelleck, Wales, England
School Affiliation: Cambridge
Best Work: *Principia Mathematica*
Best Formula: Can a set be a member of itself?
Fields of Interest: Deduction, logic
Quote: Where we validly infer one proposition from another, we do so in virtue of a relation which holds between the two propositions whether we perceive it or not

Bertrand Russell

It's rare when a philosopher becomes as well-known as Bertrand Russell, at least while he or she still lives. Russell won the 1950 Nobel prize for literature, and his outspoken pacifism, combined with first-rate mathematical and philosophical work, made Russell something of a living legend. Only Hitler finally budged him from his dedication to total pacifism.

Throughout his life he was a vocal critic of religion and politics, and publishing books like *Why I Am Not A Christian* did much to make him famous, if only as a liberal atheist. He and his wife founded the experimental Beacon Hill School.

The price of being a good teacher is that you're sometimes remembered for what your students did. Ludwig Wittgenstein proved to be one of those students. While with Russell, Wittgenstein turned to logic and explored the ways it defined language. If Russell's emphasis was on deriving mathematics from logic, Wittgenstein's work probed the relationship between logic and language. For Wittgenstein, the two were inseparable.

Russell was in the front ranks of the '60s "Ban the Bomb" movement, and he teamed with existentialist philosopher Jean-Paul Sartre to fight U.S. involvement in Vietnam. To earn his bread and butter, Russell produced several "coffee-table" books on philosophy and history, and, though popular works, they're also trenchant introductions to the history of Western philosophy.

Principia Mathematica was just the latest attempt to derive mathematics from a short list of logical axioms, and even when completed it wouldn't address the question Hilbert had posed in 1900. So again, at the mathematical congress of 1928, Hilbert restated the questions he most wanted answered, "Was mathematics complete, consistent, and decidable?"

To be complete, mathematics must be able to show whether any properly formed statement is true or false. To be consistent, it would have to be shown that mathematics could never arrive at false conclusions from valid reasoning. To be decidable, there must be a definite method for deciding the truth or falsity of any statement. With Paul Bernays, Hilbert created an axiomatic system with which to prove mathematics complete, consistent, and decidable, and Hilbert still maintained that each of these questions would be answered with a "yes."

Gödel Numbers

It's an irony of timing that Hilbert had practically just finished speaking to the 1928 congress of mathematicians when an unknown Austrian mathematician named Kurt Gödel not only answered one of Hilbert's questions but also answered *the* Hilbert question.

Gödel assaulted the challenge with the sharpest of logic and, to trap the completeness question, he used the very meta-mathematical system Hilbert and Bernays championed. On the way to the completeness answer, Gödel did what only a few mathematicians had done. He found a new kind of number.

You might have thought there could be no new numbers by the 1930s, but Gödel numbers were new. Where ordinary numbers might be said to indicate quantity or measurement, Gödel numbers represented logical sentences. A whisper of Euclid and his proof of the infinitude of primes can be heard in Gödel's reasoning, which employs the primes to create his astonishing Gödel numbers.

Gödel took every symbol used in mathematical logic and assigned it a number. Say the number assigned to the logical connective v was 3, and the numbers assigned to the letters p, q are 5, 7. Then the statement p v q has the Gödel number 5 x 3 x 7. Every statement, every formula, and every formalized rule of arithmetic could be represented by a distinct integer, its Gödel number. Any Gödel number, broken back down into its factors, could be translated into the logical sentence it stood for.

Then Gödel surrounded the completeness problem with his numbers. Gödel created a Gödel number whose logical sentence talked about itself. Gödel used his numbers to show how to create a sentence in mathematical logic that says, "I cannot be proved." In a system that's complete, every properly formed sentence could be shown to be either true or false, not both. If the sentence "I cannot be proved" is proved, then it's false, and if the sentence "I cannot be proved" is true, then it's false, for if it's true that "I cannot be proved" cannot be proved, then there is a truth that cannot be proved, which leaves the system incomplete.

What Gödel showed was striking, and not at all what Hilbert and others had been looking for. Using his new numbers, the little-known Austrian published a paper that rocked the mathematical world. With his unique system, he found a statement that could not be proved using the given postulates. He showed that even when this statement was added to the postulates, he could always construct another statement that would say, in effect, "I cannot be proved."

Not only that, Gödel's proof also struck the heart of the far-reaching answer Hilbert had sought. The Incompleteness theorem proved not only the Hilbert/Bernays system incomplete, it also showed that any system of axioms strong enough to capture arithmetic would capture unprovable statements, meaning mathematics would forever remain incomplete.

Incompleteness was startling indeed, and devastating to Hilbert's quest, but mathematics continued. Though an upper limit seemed placed on the soundness of mathematical logic, there was, after all plenty of math left to do, and there was a powerful new tool coming that would help create whole new fields of mathematics.

Kurt
Gödel

Born: 1906
Died: 1978
Hometown: Brunn (now Czechoslovakia)
School Affiliation: Institute for Advanced Studies
Best Work: *On Formally Undecidable Propositions*
Best Formula: The Incompleteness Theorem
Fields of Interest: Logic, mathematical systems
Quote: The human mind is incapable of formulating all its mathematical intuitions

By most accounts, this curious character fits the mathematical stereotype to a T. Aloof but not always serious, Gödel seemed to operate out of a distrust of authority that he shared with many others who escaped the Nazi terror. He was only 25 when he published the Incompleteness Theorem in Vienna in 1931. He left Europe to teach at the Institute, where Einstein presided.

His colleagues related the uncanny way Gödel seemed to follow their arguments to the bitter logical end before they had fully explained them. When he could be found, that is.

His acute sense of distrust caused him to make any appointments people requested, but at the appointed time and place Gödel would never show. When Einstein asked him about this, he said that only by booking a meeting could he know for sure where the people would be. It was the surest way never to see them.

There's a story of Gödel getting his American citizenship. His distrust of authority could prove disastrous for his chances, no matter about the Incompleteness Theorem. Einstein accompanied him and made him promise just to answer the questions and play the bureaucracy game. Gödel, however, could not sit still and listen to what he thought were illogical arguments. He proceeded to tell the examiner about several logical fallacies he had found in the American Constitution. Only Einstein's explanation and pleading got Gödel his citizenship.

Computers

At one time computers were people who computed. The abacus is a computer and so is the slide rule, but the person who could use them correctly was the real computer. Calculating today is only a matter of pushing the right buttons, and the slide rule and abacus are toys compared to today's desktop computers.

The abacus and slide rule weren't automatic, and very large calculations were still difficult with these tools. You may have a computer on your desk right now, or even in your lap, but the personal computer is a very recent development. The search for an infallible calculating machine is not a recent development. By the time of the Industrial Revolution all sorts of work was being done by machines, so why not calculations?

Putting It On Automatic

The slide rule was the best tool for calculating from its invention in the 1620s to the middle of the 20th century. Slide rules were better than hand calculating, but they could be hard to read. If the markings on the rule were made farther apart, the overall size of the slide rule itself would grow. In 1892, Edwin Thatcher created a slide rule that could be read accurately to five decimal places. If it had been built like an ordinary slide rule, it would have been over 30 feet long. Instead, Thatcher "chopped up" the giant slide rule and mounted the pieces around an 18-inch cylinder that looked like a rolling pin. By spinning to the appropriate scale, you could easily read the numbers.

Pascal was born around the time the slide rule was invented, and before he was 20 he had tried his hand at building a true calculating machine. It was built of meshed gears that fit in a little box. Each wheel had 10 positions to represent the digits 0 through 9. By turning the interlocking wheels, Pascal used his machine to add and subtract numbers. This is still the basic design for mechanical adding machines, though the wheels are no longer hand-cranked.

Gottfried Leibnitz built a similar machine toward the end of the 16th century, and this one was able to multiply. Precise stamped gear wheels weren't available yet, and his machine wasn't always accurate. For difficult calculations, most people used a slide rule or logarithm tables. All sorts of mathematical tables were in use, but they took time to compute and compile. A pure calculating machine seemed out of reach, but maybe a machine could be made that would produce mathematical tables.

Born: 1792
Died: 1871
Hometown: London
School Affiliation: Cambridge
Best Work: The analytic engine
Best Formula: He mechanized the difference formula:
$$y = a_N x^N + a_{N-1} x^{N-1} + \ldots + a_1 x + a^0$$
Fields of Interest: Mathematical tables, computers
Quote: I wish to God these calculations had been executed by steam

Charles Babbage

Charles Babbage needed peace and quiet to think about building his computer. He was occasionally seen chasing organ grinders and street musicians away from outside his window. He had set himself the goal of building a machine to quickly churn out accurate mathematical tables for use by bankers, scientists, and anyone who might need such a device. He had no time for street performers.

After finding errors in the mathematical tables used by British astronomer John Herschel, Babbage uttered the words quoted above. No doubt Herschel complained as well, but Babbage did something about it. Babbage focused on tables that were calculated using the method of differences, so Babbage called his machine the difference engine. He designed it by himself, but to finance it he went to the British government.

The government was good for about 17,000 pounds, and Babbage put his money where his math was with about a third again as much. Multiple designs were developed, and many of the parts were manufactured, but the difference engine was never finished. Babbage and his engineer disagreed over some money-related matter, and in 1833 the project was abandoned. It had been an expensive failure considering that in 1831 you could build a steam locomotive for a little under 800 pounds.

The failure of Babbage's computer project didn't stop him from trying again, and in 1834 Babbage began work on what he called the analytic engine. This would be a more powerful computer, one that could be "programmed" to do different tasks, any tasks, not just generate mathematical tables. Babbage worked to build the engine until his death in 1871. He had spent much of his money on his ideas, but neither engine was ever completed. (Until recently, when the difference machine was built to Babbage's specifications. It's now on display in the Science Museum in London.)

Babbage's analytic engine would be capable of being programmed to accomplish different tasks. In 1800, Frenchman Joseph-Marie Jacquard built a programmable loom that could weave intricate patterns automatically, getting its instructions from cards with holes in them. By carefully punching holes in a series of cards, the loom would allow needles with thread to enter the pattern only at a certain places. No holes, no thread. Although making the cards themselves took time, once the cards were in shape the machine could accurately reproduce that pattern over and over again.

Babbage thought the same method could be employed to turn gear wheels and calculate mathematical tables. Depending on how the cards were punched, the machine would add, subtract, or do whatever else you needed it to do. This analytic engine of Babbage's drew support from many influential people, but he never completed it. Lord Byron's daughter, the Lady Lovelace, was a great supporter of Babbage's computer. She thought of the analytic engine as very much like the Jacquard loom, except instead of fabric it would "weave" mathematical formulas.

The beginnings of modern computers can be found in the efforts of code breakers of WWII. Among those, Alan Turing contributed significantly to the construction of the first computers and a different approach to the idea of a machine to "weave" formulas according to how it was programmed. Turing thought the computer might become a real thinking machine. One of his ideas was to equip a machine with "television cameras, microphones, loudspeakers, wheels, and handling servo-mechanisms" as well as some sort of "electronic brain ... [that would] roam the countryside ... finding things out for itself."

*Alan
Turing*

Born: 1912
Died: 1954
Hometown: London
School Affiliation: Trinity College, Cambridge
Best Work: "Computable Numbers"
Best Formula: Any method of human computation can be translated to a method of machine computation

Fields of Interest: Computers, artificial intelligence, cryptoanalytic activity
Quote: Dip the apple in the brew, let the sleeping death seep through

Alan Turing was raised in England by his mother and aunt. As a boy, he demonstrated an interest in science and numbers, but he was most famous for messy ink spots on his clothes. He was rarely seen with his shirt tail tucked in, and at public school his soccer coach said that "Alan spent too much time watching the daisies grow."

Instead of smoking up the house with his experiments, Alan would let smoking pots of chemicals cool down on his bedroom window ledge. By all accounts he was a peculiar boy with a talent for numbers, and he was always inventing things. Alan and his friend David Champernoune developed "round-the-house" chess, which did away with chess clocks. Once white moved, that player got up and ran around the house. Black had to move before white returned to the board, then black ran around the house, etc. The faster you ran the less time for your opponent to plan his move.

Turing wrote "On Computable Numbers with An Application To Entscheidungs' Problem," and took part in an exchange program to study for a year with Alonzo Church at Princeton. In the university metal shop, Turing wound wire to create relays and built a simple electric multiplier. Turing returned to Cambridge University in 1938 with his multiplier under his arm, and the idea for a machine that could be programmed to solve lots of different questions.

Turing joined the code-breaking arm of the military and immediately contributed concrete procedures and methods that saved time deciphering the German code—and thereby saved lives. Once the code breakers got underway, Turing was eased out of the administrative and managerial operations of code-breaking headquarters Hut 8. Great minds aren't always great managers, and Turing's supervisors knew it. He was made chief consultant, out of the day-to-day loop, and soon Hut 8's memos and directives lost Turing's tell-tale ink-splotches.

After the war, Turing continued to work on the development of the computer. He was also a part-time marathon runner and a Cambridge professor. He spoke about computers on the BBC radio—not only the potential for computing machines but also the possibility of creating thinking machines. He was also arrested for being homosexual.

Turing was a homosexual when homosexuality was against the law. The British government put Turing on a regiment of hormone therapy to "cure" him. The same government that recruited him to break the Nazi code and praised the Cambridge genius made Turing an outcast.

In effect, Turing had been chemically castrated because of his sexual orientation, and so in June 1954 he coated an apple with cyanide and ate it until he died. His brilliant and far-reaching notions of a "thinking" machine became crucial to the development of the computer, and this was work he accomplished while still a young man.

The Enigma

The Germans used different forms of the Enigma encoding machine to communicate with their spies, ships, and U-boats. The Enigma was originally designed for use with international business communications, but the Nazi code teams added many devious flourishes to this mechanical coder, and it was crucial to the Allies that the Enigma be broken.

The mechanism is reminiscent of Jefferson's wooden-wheeled coder, but these were metal wheels with meshing teeth. German letters covered the edges of the wheels like a substitution alphabet, and every time a letter was typed the code wheels slid again, reshifting the substitution alphabet. The whole thing was dependent on initial settings that were agreed on by sender and sendee, and changed frequently to thwart deciphering.

An exterior plugboard to the Enigma could be used to transform 10 letters in one of 150,738,274,937,250 possible ways. There were lots of other tricks the Germans used to code their messages, and they continued to add flourishes to the Enigma that made its code almost impenetrable.

Alan Turing and the British code breakers attacked the Enigma and developed a form of mathematical "guessing" to aid in quick solutions to coded messages. This "weight-of-evidence" procedure, later to be called sequential analysis, might save an hour or two when breaking a code, and an hour was the time it took a U-boat to gain 6 miles on a convoy.

It was in part the final success of the code breakers of WWII that convinced governments of the usefulness of expensive computing machines. Alan

Turing had been at the center of that success, and he continued to pursue his ideas of computing machines after the war ended.

Turing Machines and Turing Tests

After the war, Alan Turing showed how simple machines with very limited capability could be made to solve any complicated problem as long as the problem could be formulated in mathematical terms. The Turing Machine was only theoretical, but it demonstrated how very complicated problems could be solved by means of a few simple procedures, and it made people wonder whether the complicated process of thinking could be mechanized.

Turing thought so, and even devised a test for "artificial intelligence." If after communicating with a person or a machine only through a keyboard and monitor you were unable to tell whether you were communicating with a person or a computer—and you were communicating with a computer—then that computer would have demonstrated "artificial intelligence." Today competitions are held to see how close different programs can come to passing the Turing test.

Fractals

I n the 19th century, logical systems and mathematical philosophy were at the center ring of the mathematical world, and mathematics itself became an object of mathematical thought. Few mathematicians were looking to the "real" world for their ideas, and even when mathematics was employed to explain the physics of relativity it was absorbed with symbols on a blackboard. The mathematicians were scribbling waves of symbols that seemed only vaguely related to the world.

One mathematical explorer did look to nature as his subject. When he looked at things in the world, he thought he saw a new type of mathematics, maybe even a new way of looking at the structure of nature. Even more strange was that the mathematics Benoit Mandelbröt saw he saw everywhere he looked. He saw it in clouds and mountains, but he also saw it in the rise and fall of cotton prices. He heard it on noisy telephone lines, and he came face to face with it when he measured the coastline of Britain.

Triangles Again, and Snowflakes

Benoit Mandelbröt deserves the credit for making "fractals" a commonplace word, but others before him had seen glimmers of fractals. As if we haven't leaned on the poor triangle enough already, one of the early glimpses of fractals begins as a special triangle, one that doesn't stay a triangle for very long.

It's easy to make one of these triangles. Start with an equilateral triangle, one whose three sides measure the same. In the middle of each side, construct another triangle that's one-third the size of the original. The resulting shape is the familiar Star of David. Now there are 12 sides to the figure. In the middle of each of these sides, draw another triangle, one-third the size, and place it on each of the 12 sides. Now the figure is beginning to resemble a snowflake. Apply the same simple rule as many times as you like and the snowflake will grow more intricate.

Known as the Koch snowflake in honor of its creator, Swedish mathematician Helge von Koch, it was a puzzlement in 1904 when first presented. There are some peculiar aspects to this figure that bear on fractals. There is no theoretical limit to how many times you perform the snowflake transformation on the shape. In fact, you could keep doing it forever. Mathematically, at least, that makes the border of a Koch Snowflake a potentially infinite line, but the area inside is finite! In fact, a circle drawn to include the original triangle will always include a Koch snowflake made from that triangle, no matter how many times you divide the line with another triangle. The figure will change rapidly into a complicated snowflake, yet you can always find triangles on the edge.

The master of infinity, Georg Cantor, created a similarly unusual pattern, but not by adding on to a triangle. Cantor created his "dust" by removing parts of a shape. Brutally simple but perplexing nonetheless, the Cantor Set is easy to create. Start with a line and remove the middle third. You'll be left with two lines, to which you apply the same procedure and remove the middle third of each line. Continue as long as you like and you'll be left with more and more (yet smaller and smaller) lines.

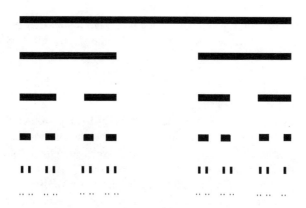

The Cantor Set

If you carry Cantor's procedure far enough, tiny specks will be all that will remain; this is Cantor's "dust." Though you can always continue to erase part of them, they don't disappear. If we apply Cantor's rule forever, we'll end up with an infinite number of lines—marks, really—whose total length adds up to zero.

Do You Hear What I Hear?

These and other constructions perplexed and befuddled many 19th-century mathematicians, but Benoit Mandelbröt thought he saw in the Cantor Set what he had been hearing over telephone transmission lines. The Cantor set was kind of messy, and so were the transmissions IBM was sending over telephone lines. Mandelbröt had been hired to clean the noise out of these transmission lines, but he found that some noise just couldn't be removed, like Cantor's dust. What struck Mandelbröt was not that the Cantor Set resembled the pattern of noise heard over transmission wires, but that it so perfectly resembled the noise.

The electronic noise that Mandelbröt analyzed seemed to come in a patternless, random way. Boosting the power of the transmission got rid of much of the noise, but never all. Though engineers had begun to detect patterns in the random noise of telephone transmissions, Mandelbröt found a way to analyze it and confirmed his theory that the noise not only came in clusters, it came in a pattern.

Mandelbröt divided the time of transmission into different slices. On an hour-to-hour basis, he found that an hour would tick off without any noise, then an hour would have noise. The same thing held if the transmissions were divided into 20-minute slices: Some would contain errors, others wouldn't. In five-minute slices, Mandelbröt found the same thing. No matter how he divided it, the transmissions would have some noise and some clean periods. Some noise was going to be part of the transmissions.

The Map Is Not The Territory

If you want to know the length of the coastline of Britain, you could go to an almanac or atlas and look it up. If you couldn't locate the information, you could go measure it yourself. You could rent a car, drive around the island, and read the distance off the odometer, but your answer would be inaccurate.

For more accuracy, you might measure it with a yardstick. This would take much longer, and you'd still have to make certain compromises. A yardstick can't be bent around every rock on the coastline, and perfectly straight yards can't measure the smallest jagged edges where the water meets the land. A ruler would be even more time-consuming, but a ruler would capture much more detail than a yardstick. An even better figure for the length of the coastline could be obtained using a one-inch ruler.

The coastline of Britain measures longer when measured with a ruler than when measured with a yardstick. Imagine using a grain of rice as the measuring unit. If the length of the coastline gets bigger depending on what measuring stick we use, then how can we say how long it is? What can we learn by measuring the coastline if not its length? In the early '60s, Benoit Mandelbröt published a paper titled "How Long Is the Coast of Britain?"

Mandelbröt thought that though the length of coastline changed depending on the measuring stick, something remained the same. No matter what unit of measure you used, the essential shape of the coastline didn't change. The rocky coastline has a texture that shows up whatever measuring stick you use—even if you used the grain-of-rice measuring stick or took measurements from an orbiting satellite. Measuring Britain's coastline didn't answer the question "how long," but it seemed to provide clues to another question: "What is the shape of the coastline of England?"

Fractional Dimensions

If you measure a five-meter line with a meter stick, the length will be 5. The same line measured by units of 10 centimeters will yield a length of 50. The ratio between the measuring units remains the same: 50 units of 1/10, five units of one, or 50:5, which reduces to 10:1. This ratio can be written as an exponent, 10^1, and it's the exponent we're interested in. The exponent, 1, is the dimension of the object—in our case a straight line, which is a one-dimensional object.

The same reasoning works for two-dimensional objects, like squares. A square with sides of five meters could be measured by smaller squares of one-meter length. It would take 25 one-meter squares to fill a bigger square with sides five meters long. It'd take 2500 squares with sides of 10 centimeters to fill the same five-meter square. Here the ratios are 2500:25, which is 100:1, or 10^2. The exponent is 2 because squares are two-dimensional objects.

You may expect that this ratio will hold for three-dimensional objects, and it does. A cube of five meters can be filled by 125 one-meter cubes, or 125,000 10-centimeter cubes, and this ratio, 125,000/125 reduces to 1000, or 10^3. The exponent represents the three-dimensional aspect of the cube. By examining the ratio between the number of units required to fill a shape and the size of those units, the dimension of the shape is revealed.

So what is the shape of the coastline of Britain? The method above was developed by the German mathematician Felix Hausdorff and when applied to the coastline of Britain. The fractional Hausdorff dimension of the coastline is ≈1.33. Fractional dimensions, or fractals (Mandelbröt coined the term), describe the roughness of the rocky coastline or the fine roughness of the Sahara. No matter what magnification you use to examine a coastline, its shape texture remains the same.

Benoit Mandelbröt

Born: 1924
Died: still with us
Hometown: Warsaw, Poland
School Affiliation: Paris Ecole Polytechnique, Harvard University
Best Work: *The Fractal Geometry of Nature*
Best Formula: D (fractal dimension) = logN/log(1/r)
Fields of Interest: Geometry, fractals
Quote: To see is to believe

For years Benoit Mandelbröt raved about his fractals to an unbelieving or uninterested community of mathematicians. After all, by the early '60s they were well-accustomed to pushing symbols all over blackboards, and looking at mountainscapes, snowflakes, and cotton prices seemed far removed from the real work of mathematics.

Mandelbröt's resume is varied, to say the least. He worked with IBM, studied the rise and fall of cotton prices, and held several other positions, making him something of a jack of all trades. So to make his ideas heard, Mandelbröt beat his own drum (some say his chest), and made it his mission to get the subject of fractals into the mainstream.

He plotted the Mandelbröt set, a fractal consisting of all the complex numbers that, when plugged into a formula, don't grow infinitely large but stay small. When plotted, the Mandelbröt set looks something like a self-similar bug, with more and more detail at every magnification. And he did create, name, and promote the science of fractals, though jealousy prevented many from swallowing Mandelbröt's story of how *he* discovered the relations; how *he* created the formulas; how *he* built the Mandelbröt set, probably the most-reproduced result of the new world of fractals.

Self-Similarity

Rocks look like the mountains they're a part of. The shape of the whole mountain is also the shape of its smaller pieces. Twigs grow like the trunks of the trees they branch off of, and clouds look pretty much the same whether they're seen from a hillside or an airplane. We can think of a fractal as any complex geometric shape that exhibits the property of self-similarity. Self-similar objects have components that resemble the whole. A fractal object tends to look the same no matter what magnification you use to look at it. Fractal objects remain invariant under changes of scale. This invariance under change of size is known as scaling symmetry.

The interesting thing about fractals is that they can be used to represent many irregularly shaped or spatially non-uniform phenomena in nature, and this is something that tired old Euclidian geometry can't do. Bridges, buildings, and other man-made objects are represented by the Euclidian system, but for nature the fractal is better. The basic characteristics of a fractal object will remain the same no matter how big it's enlarged or how small it's shrunk, and these unchanging characteristics are its fractal dimensions. A fractal is self-similar and demonstrates a non-integer dimensionality.

There is a self-similarity to the shapes of these natural object that can be accurately described by fractals. This new branch of mathematics is being employed successfully in the plotting of galaxies, the study of fluid turbulence, and, most amazingly, in computer graphics. Not only do fractal equations produce superior representations of natural objects, the nature of a fractal's reiteration process—the applying of a simple rule over and over again—is what a computer does best.

When a number is arrived at by a formula, fed back into the formula to get another number, and that subsequent number fed back into the formula, you may also get some rocking music. Not everyone, perhaps, will recognize feedback as music, but in the hands of a virtuoso original like Jimi Hendrix I think it's hard to deny.

Feedback is created when the sound from an amplifier's speakers is picked up by the same microphone that produced the sound. It doesn't take long for the signal to feed back many times through the microphone/speaker loop, creating an endless chaotic spiral. Chaotic unless it comes from Hendrix or someone else capable of containing it. "The Star Spangled Banner" with feedback is the only version I can sing along with.

Could this be the real language of nature? Compared to fractals, standard geometry fails miserably to capture the essence of the natural world. Fractals are used in Hollywood to create animations of mountains, storms, and fire. Fractals on the computer are used to create lots of special effects we see in movies today. As Mandelbröt said early on, "Clouds are not squares, and waves are not curves." He taught mathematicians once again that seeing is believing.

Chaos

Most of the time just missing the bus or burning the sauce is enough to make a day chaotic. Really chaotic days are hectic, frenzied, and crazy—a complicated mess. Mathematicians think of chaos as something more than a complicated mess. To them, chaos is utter, complete, and impenetrable disorder and confusion. Chaotic events are totally random and completely unpredictable. To have a mathematically chaotic day would be an unusual day indeed.

Some of the descriptions of chaos make it sound like the end of the world, but there are plenty of everyday examples of chaos. The air that flows over a jet airplane wing is an example of chaotic behavior. So is the growth of animal populations, the rise and fall of the stock market, and the unpredictable nature of the weather. A thin stream of water from the kitchen tap will appear to be one thin stream, and, as you turn the tap open wider, the shape of the stream will appear to change smoothly. But when the tap is opened wide enough, the stream turns turbulent, and that turbulence is chaotic.

Just like the weather, most people who talked about chaotic processes never did anything about them. After all, what could be done? What is there to understand in a chaotic system, and if some kind of order is found within that system is the system still chaotic? This new science of chaos, like fractal geometry, was born out of physical observations of chaos. People met chaos when they looked at thunderheads, listened to feedback, or thought about butterflies.

Three Body Problems

Newtonian physics captured the essence of natural phenomena in simple formulas. The formulas were simple because they described simple, isolated physical problems. Simple problems can be formed in precise terms, so you can figure out the motions of the moon and predict its location using a simple formula. All you have to do is find out where the moon is now, apply the formula for orbiting bodies, and make predictions about where the moon will be next week or next year. Simple concepts like orbiting can be explained with simple formulas.

Little differences might throw your answer off. Maybe you didn't get a precise enough measurement of the moon's original location. That error would throw off your prediction, more or less depending on how large an error you made. Still, the formula works, and today the motion of the moon is essentially predictable. By extending the system to include all the other bodies in the solar system, the galaxy, and the universe, it would seem that you could make accurate predictions about them all. Trouble is, your predictions go south before you can even add a third body to the picture.

Add the sun into the system and the numbers go off. You can adjust for some variance, but you won't have the predictive success you did with just two bodies. When it comes to three bodies or more, only approximate answers are possible. No matter how precise your information, lots of little effects still add up to throw the numbers off the mark. Thinking all discrepancies from prediction could be traced to some observational error, scientists dismiss these deviations from prediction as "noise."

Pinning down noise is hard. When it comes to complicated problems like three bodies, no matter how much information you plug in you'll always get some noise in the equation. The more complicated the system, the more noise. Our weather system is a very complicated system, though in essence it's just the sun's energy heating our atmosphere. Red sunsets and violent cyclones are just features of the same complicated system, but to predict their behavior you'd need lots of precise information indeed.

Studying chaos requires a lot of calculations. An iterated program is one that arrives at a value and then plugs that value back into the formula. It's no wonder you can find chaos on a computer—they're built especially for it.

A Tempest on a Terminal

Computers are designed to do massive calculations over and over again, and in 1961 a meteorologist named Edward Lorentz was using a computer to make the weather. Of course, this was no real weather. Instead of rain and hail, all Lorentz really produced was zeros and ones, but to him they represented massive storm systems, high and low pressure pockets, and wind, all moving across the page. He didn't even have a terminal, but waited patiently while his printer churned out pages of zeros and ones, pages of storms and calm, pages of weather, and, finally, pages of chaos.

One day Lorentz was looking at the zeros and ones, flipping through pages of printout weather. He saw a very interesting piece of weather and decided to run the program again. The computer was programmed to calculate numbers to six decimals, but when it printed out the results only the first three decimals were printed. Of course, this difference is very small indeed, hardly worth worrying over, so Lorentz plugged in the numbers he did have, with only the first three decimals. It didn't occur to him that the small difference in the decimal value of the two sets of numbers might be enough to affect the interesting storms he'd been watching.

The weather Lorentz saw when he ran his program wasn't just different from the first weather, it was completely different. The tiny differences that Lorentz wasn't able to accommodate had created a weather system that was nothing like the weather he was trying to duplicate. It was surprising, but the tiny differences accumulated over time, and, after a while, the second weather system was totally distinguishable from the first. The beginning numbers that Lorentz plugged into his weather formula were key to the weather's ultimate behavior. The weather system demonstrated a "sensitive dependence on initial conditions." Though he didn't know it, Lorentz had stumbled onto one of the features of a chaotic system.

The Butterfly Effect

Even if you only miss tiny factors when you try to duplicate a system's initial conditions, repeated calculations magnify small factors. Meteorologists had already noticed what Lorentz saw in his computer weather. They knew that no matter how much detailed information they plugged into their weather-predicting programs, they could never accommodate enough information to represent initial conditions. They called it the butterfly effect.

It was as if the flapping of a butterfly's wing in China could contribute to the power of a Gulf Coast hurricane. Weather systems were so complicated you could never have enough information to make long-range predictions about them. There was no way to get enough detailed information for the formula to work. Some small factors would escape you, and over time they'd throw your predictions off. It's this sensitive dependence to initial conditions that makes complicated systems like the weather impossible to predict. The outcome at the end can be very different when even small changes are made in the initial conditions.

Strange Attractors

We know some infinite sequences converge to a value, and some of the iterated formulas used to predict weather or animal populations tend toward certain values. Unlike a truly converging sequence, these equations zero in on a certain value, or several values. These values are said to be attractors.

In the simple geometrical sequence, $1 + 4 + 9 + 16 + ...$, the numbers will get farther and farther away from zero. In this case, the values of the formula act as if they are repelled by zero. Unlike convergent sequences, iterated formulas that have strange attractors show a self-similar pattern of numbers that never repeat exactly, but are attracted to a value or several values.

When Lorentz plotted the results of his weather program, he got a weird figure-eight picture of two swirls, like butterfly wings. The chaotic weather system on Lorentz's computer had a kind of order to it when graphed. Though the figure kept tracing the same shape over and over again, it did so with slightly different values, meaning no value was ever repeated. Though chaotic, the Lorentz butterfly—as the graph came to be called—showed that even a chaotic system has a kind of order to it.

Order in chaotic systems began to be observed in many different fields. Epidemiologist William Schaffer of the University of Arizona thinks he has developed a chaotic system that mimics the spread of epidemics. Jack Wisdom of M.I.T. thinks the chaotic nature of the asteroid belt might account for the behavior of the asteroids that somehow leave the belt to bombard other planets.

So there is a kind of order to chaotic behavior. The values of a chaotic system may never repeat exactly, but they seem to hover around a strange attractor or two—or many. Chaos can be found just about anywhere, and as this new field of science expands we may come to think of the "noise" in our formulas as just as important as the formulas themselves.

Problems

Mathematics does things, and for the most part what it does is solve problems. Some problems are more practical than others, like figuring out your taxes, and a good engineer can excel in that field without a knowledge of Cantorian Set theory. Tax accountants have more pressing problems than considering pure number theory, though occasionally a deduction may also be a prime number.

People often make a distinction between "pure" and "applied" mathematics, though what is pure mathematics at one time may be perfectly suited to some application years later. Lots of non-practical mathematical ideas are later dusted off and put into service when their usefulness becomes apparent. In many ways, mathematics doesn't exist outside of some interpretation or application. Even the obvious, "2 + 2 = 4," needs some kind of setting in which to be "true." Before you agree with the formula, you need to make sure that it doesn't refer to Alka-Seltzer and Pepsi Cola. I don't know what you end up with when you add two boxes of Alka-Seltzer to two liters of Pepsi, but I doubt it's four of anything.

There's a bias against applied math. For some lofty mathematicians, once something is proved the game is over; if someone finds value in the work, if the math is useful, so be it, but once it's been demonstrated to be "true" the "pure" mathematician has other things to do. The tussle between applied and pure mathematics was summed up concisely in *A Mathematician's Apology* by G.H. Hardy. To be a pure mathematician—as Hardy certainly considered himself—a person's entire body of work must be useless. The game is the hunt for mathematical truth, not engineering techniques.

It's no wonder, then, that so many mathematicians tried to "square the circle," or trisect the angle. Not being able to do something has little bearing on mathematical reality. For thousands of years, only the first few perfect numbers were known, but this in no way meant that there weren't other perfect numbers. Still, mathematical truth must be shown, or demonstrated. A million examples that fail to prove a formula mean little. For mathematical reality, you have to prove something is or isn't so. Until that's done, mathematicians consider the question unanswered.

The Four-color Map Problem

The four-color map problem was unanswered until 1976, though it's easy to state the problem. If you want to make money selling maps, you cut costs where you can. One way is to print the maps using as few colors as possible. Of course, you still want to produce useful maps, so bordering countries must be different colors. For a map with only two countries, you'll need only two different colors. A map with three countries might look like a peace symbol, and would require at least three different colors to make it useful.

Obviously we'll need to make maps with more than three countries. To refine the problem, countries aren't allowed to have separate colonies, and where countries meet at a sharp point, they don't border each other. Without this last definition, countries shaped like slices of a pizza would all need a separate color.

So what's the fewest number of colors you can get away with? In 1852, Francis Guthrie, a graduate student, mailed a letter home to his younger brother. In that letter, he wondered whether four colors were enough for any map. Lots of people tried to prove that four was enough. The problem is simple to state and simple to diagram.

With just a few lines you can draw an arrangement of countries that requires four colors, like so:

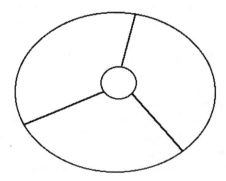

This map requires four colors

When De Morgan wasn't busy doing logic, he thought about the problem. His attempts seemed to indicate that you just couldn't draw a map that would require five colors. It certainly looked like four colors was enough, but that's not proof.

The four-color map problem was solved in 1976 by Kenneth Apel and Wolfgang Haken, taking over 1,000 hours of computer time. In fact, the whole proof is more than 100 pages long, and depends crucially on the results of computer calculations. The computer obtained results used in the proof that would be practically impossible to check by hand, so the proof by computer can only be checked by computer. Maybe it's only a proof to the computer.

Apel and Haken certainly acknowledge the importance of the computer in their results. The computer was, in fact, essential to the proof. Acceptance of this solution to the four-color problem is granted by most and conceded by others. A shorter, more elegant solution is desirable, but is it necessary?

The Goldbach Conjecture

Another letter, this time between well-established mathematicians, included what is called the Goldbach Conjecture. Euler was the first to hear of the problem when he received a letter from his Prussian mathematician friend Christian Goldbach. What Euler read was a simple idea about the structure of numbers and how they're formed out of the primes.

Simply put, Goldbach said that every even number greater than 4 can be expressed as the sum of odd primes. It's easy to find a few even numbers that show the conjecture valid: $8 = 7 + 1$; $24 = 19 + 5$; $88 = 71 + 17$ or $83 + 5$. Showing examples of the conjecture tends to make us think the conjecture is true, but by now you know that that's not enough for mathematics.

Another version, called the "weak" Goldbach conjecture, says that every odd number larger than 7 can be expressed as the sum of three odd primes. Some examples are: $9 = 3 + 3 + 3$; $25 = 5 + 7 + 13$; $95 = 17 + 31 + 47$. Even this weak conjecture has refused to be proved. In 1937, the Russian mathematician Ivan M. Vinogradov demonstrated that the "weak" conjecture is true for all "sufficiently large" numbers. Though Vinogradov wasn't able to say what a sufficiently large number was, 20 years later his student K.V. Borodzin did. For numbers larger than 3 cubed to the 15th, the weak conjecture is true. See if your calculator can display this number.

The strong conjecture yielded to similar attack, this time in 1966 by Chinese mathematician Chen Jing-run. He showed that, for big enough numbers, those numbers can be expressed as the sum of a prime plus a number that's either prime or the sum of two primes. This is nice to know, but leaves the question of Goldbach's conjecture unanswered. Both strong and weak versions have been shown to hold for numbers up to a million or more, but examples aren't proof.

Putting Math in its Place

Even among mathematicians it's not agreed what emphasis is appropriate for the study of mathematics. Mathematics can be the "Queen of the Sciences" or just modern physics' stepchild. For Plato, mathematics represented eternal truth, and the Pythagoreans developed a mystical appreciation for the beauty of math. Does current mathematical thought preserve any of these aspects of eternal truth?

Much of Hilbert's work dealt with logical foundations and systems for mathematics. Mathematical systems are like games with intricate rules and restrictions. Yet, as Gödel showed, no formal system can completely capture the truths of mathematics. Still, mathematicians can't just create any system at all. There are essential internal relationships that constitute a "mathematical" structure, no matter how much of a game it may seem. The proudly "pure" mathematician G.H.Hardy said of primes, "317 is a prime, not because we think so, or because our minds are shaped in one way rather than another, but because it is so, because mathematical reality is built that way." For at least one great mathematician early this century, mathematical equations expressed a thought of God.

Srinivasa Ramanujun

Born: 1887
Died: 1920
Hometown: Madras, India
School Affiliation: Cambridge
Best Work: letters to G.W. Hardy
Best Formula:
$$\frac{1}{\pi} = \frac{\sqrt{8}}{9801} \sum_{n=0}^{\infty} \frac{(4n)!\,[1103 + 26390n]}{(n!)^4\,(396^{4n})}$$
Fields of Interest: Number theory, partitions
Quote: An equation for me has no meaning unless it expresses a thought of God

Over 100 years have passed since Ramanujun was born to a poor family in a small town in India. As a child, Ramanujun poured his energy into his fascination with numbers. Day and night he scribbled his formulas on a small slate, erasing with his elbow so he wouldn't have to put the chalk down.

Driven as he was, Ramanujun completely neglected any other studies. The Indian universities recognized his mathematical prowess, but he was so deficient in everything else that he failed repeatedly. When he was 25, he took a job as a clerk.

But Ramanujan's spirit couldn't be denied. He continued to work on his mathematical ideas. When paper was short, Ramanujun would use another color ink and fill in the spaces of already-filled pages. And there was always his furious clacking of chalk and slate.

With what precious paper he could find, Ramanujun sent desperate letters to some of the great mathematical minds of the age, begging them to consider his formulas. Though a few turned him down, most never even acknowledged this strange Indian clerk. Then, in 1913, Cambridge mathematician G.H. Hardy opened the letter that would change his life. The unknown Indian clerk's letter included several original formulas, the beauty of which struck Hardy at once. Hardy sent for Ramanujun.

In Cambridge Ramanujun thrived, at least mathematically. He lived for the discussions with fellow mathematicians, and developed a close bond with Hardy. Together they published several important papers. But the English climate was unhealthy for Ramanujun, and his religious dietary restrictions were hard to follow in WWI-era England.

The longer he stayed in England the sicker he would become, and Hardy knew it. In 1919, seriously ill, Ramanujun returned to India as a famed mathematician. Ramanujun died that April at 32. He produced three cryptic notebooks that contained almost 4,000 formulas covering his short life's work, much of which is still being studied.

Abacus, Adding Machine, Apple

Lancelot Hogben notes that, "... about four thousand years elapsed from the time men could reckon the time of the next solar eclipse, and when they could reckon how much iron is present in the sun." Other developments don't take as long. For instance, in 1974 I was a freshman in college and I used a fairly nice slide rule. During my second year, all slide rules were sold out of a bargain bin—at fire-sale prices. Four years later, every student had a calculator.

Our bank cards reduce our finances to an ID number and a security code, large sums of money exist only as electric charges in some computer system, and, as scientists get closer to deciphering the genetic code, the awesome blueprint for human life may be exposed as a complicated catalogue of numbers.

In this way, the Pythagoreans may ultimately be right. Maybe everything really is number.

Suggested Reading

Some good books about mathematical ideas:

William Warren Bartley III. *Lewis Carroll's Symbolic Logic.* Clarkson N. Potter Inc., 1977.
This fun book includes a reproduction of Carroll's game of logic.

Peter Beckmann. *A History of Pi.* Dorset Press, 1971.
The history of pi told in detail, with a special emphasis on clashes between science and politics.

E.T. Bell. *Men of Mathematics.* Simon & Schuster, 1937.
A good introduction to the lives of famous mathematicians.

Philip J. Davis and Reuben Hersh. *The Mathematical Experience.* Birkhauser, 1981.
This book covers lots of mathematical topics in essays of various length.

George Gamow. *One, two, three ... Infinity: Facts and Speculations of Science.* Bantam, 1972 (orig. pub 1947).
Still a great book about science and numbers, with good sections on relativity and topology.

Lancelot Hogben. *Mathematics for the Millions.* FRS, Norton + Company, 1937.
A venerable classic that teaches everything from arithmetic to calculus.

George Polya. *How To Solve It.* Princeton University Press, 1945.
Unusual and entertaining approaches to solving problems of all kinds.

Constance Reid. *From Zero to Infinity.* USA M.A.A, Inc., 1992.
In this book, Chapter One is about 1, Chapter Two is about 2, and so on to infinity!

Index

INDEX